POLLUTION

OPPOSING VIEWPOINTS®

OTHER BOOKS OF RELATED INTEREST

OPPOSING VIEWPOINTS SERIES

The Environment
The Environmental Crisis
Global Resources
Global Warming
Population
Water

CURRENT CONTROVERSIES SERIES

Conserving the Environment
Energy Alternatives
Garbage and Waste
Pollution

AT ISSUE SERIES

Environmental Justice
Rain Forests

POLLUTION

OPPOSING VIEWPOINTS®

Tamara L. Roleff, Book Editor

David L. Bender, Publisher
Bruno Leone, Executive Editor
Bonnie Szumski, Editorial Director
David M. Haugen, Managing Editor

OPPOSING
VIEWPOINTS®
SERIES

Greenhaven Press, Inc., San Diego, California

Cover photo: Photodisc

Library of Congress Cataloging-in-Publication Data

Roleff, Tamara L., 1959–
    Pollution : opposing viewpoints / Tamara L. Roleff, book editor.
        p.   cm. — (Opposing viewpoints series)
    Includes bibliographical references and index.
    ISBN 0-7377-0135-8 (lib. bdg. : alk. paper). —
    ISBN 0-7377-0134-X (pbk. : alk. paper)
    1. Pollution. 2. Pollution prevention. I. Title.
    TD176.7.R65 2000
    363.738—dc21
                                                                            99-17677
                                                                            CIP

Greenhaven Press, Inc., P.O. Box 289009
San Diego, CA 92198-9009

"CONGRESS SHALL MAKE NO LAW...ABRIDGING THE FREEDOM OF SPEECH, OR OF THE PRESS."

First *Amendment* to the U.S. Constitution

The basic foundation of our democracy is the First Amendment guarantee of freedom of expression. The Opposing Viewpoints Series is dedicated to the concept of this basic freedom and the idea that it is more important to practice it than to enshrine it.

CONTENTS

WHY CONSIDER OPPOSING VIEWPOINTS?

"The only way in which a human being can make some approach to knowing the whole of a subject is by hearing what can be said about it by persons of every variety of opinion and studying all modes in which it can be looked at by every character of mind. No wise man ever acquired his wisdom in any mode but this."

John Stuart Mill

In our media-intensive culture it is not difficult to find differing opinions. Thousands of newspapers and magazines and dozens of radio and television talk shows resound with differing points of view. The difficulty lies in deciding which opinion to agree with and which "experts" seem the most credible. The more inundated we become with differing opinions and claims, the more essential it is to hone critical reading and thinking skills to evaluate these ideas. Opposing Viewpoints books address this problem directly by presenting stimulating debates that can be used to enhance and teach these skills. The varied opinions contained in each book examine many different aspects of a single issue. While examining these conveniently edited opposing views, readers can develop critical thinking skills such as the ability to compare and contrast authors' credibility, facts, argumentation styles, use of persuasive techniques, and other stylistic tools. In short, the Opposing Viewpoints Series is an ideal way to attain the higher-level thinking and reading skills so essential in a culture of diverse and contradictory opinions.

In addition to providing a tool for critical thinking, Opposing Viewpoints books challenge readers to question their own strongly held opinions and assumptions. Most people form their opinions on the basis of upbringing, peer pressure, and personal, cultural, or professional bias. By reading carefully balanced opposing views, readers must directly confront new ideas as well as the opinions of those with whom they disagree. This is not to simplistically argue that everyone who reads opposing views will—or should—change his or her opinion. Instead, the series enhances readers' understanding of their own views by encouraging confrontation with opposing ideas. Careful examination of others' views can lead to the readers' understanding of the logical inconsistencies in their own opinions, perspective on why they hold an opinion, and the consideration of the possibility that their opinion requires further evaluation.

EVALUATING OTHER OPINIONS

To ensure that this type of examination occurs, Opposing Viewpoints books present all types of opinions. Prominent spokespeople on different sides of each issue as well as well-known professionals from many disciplines challenge the reader. An additional goal of the series is to provide a forum for other, less known, or even unpopular viewpoints. The opinion of an ordinary person who has had to make the decision to cut off life support from a terminally ill relative, for example, may be just as valuable and provide just as much insight as a medical ethicist's professional opinion. The editors have two additional purposes in including these less known views. One, the editors encourage readers to respect others' opinions—even when not enhanced by professional credibility. It is only by reading or listening to and objectively evaluating others' ideas that one can determine whether they are worthy of consideration. Two, the inclusion of such viewpoints encourages the important critical thinking skill of objectively evaluating an author's credentials and bias. This evaluation will illuminate an author's reasons for taking a particular stance on an issue and will aid in readers' evaluation of the author's ideas.

As series editors of the Opposing Viewpoints Series, it is our hope that these books will give readers a deeper understanding of the issues debated and an appreciation of the complexity of even seemingly simple issues when good and honest people disagree. This awareness is particularly important in a democratic society such as ours in which people enter into public debate to determine the common good. Those with whom one disagrees should not be regarded as enemies but rather as people whose views deserve careful examination and may shed light on one's own.

Thomas Jefferson once said that "difference of opinion leads to inquiry, and inquiry to truth." Jefferson, a broadly educated man, argued that "if a nation expects to be ignorant and free . . . it expects what never was and never will be." As individuals and as a nation, it is imperative that we consider the opinions of others and examine them with skill and discernment. The Opposing Viewpoints Series is intended to help readers achieve this goal.

David L. Bender & Bruno Leone,
Series Editors

INTRODUCTION

"[Effluent] trading is an innovative way for community stakeholders ... to develop more 'common sense' solutions to water quality problems in their watersheds."

Environmental Protection Agency Office of Water, 1998

"[Effluent trading] does not provide critically important safeguards to ensure that the trading of pollution credits will help meet the Clean Water Act's goals of zero discharge."

National Wildlife Federation, 1997

In 1969, an ember from a passing railcar landed on an oil slick in the heavily polluted Cuyahoga River, sparking a blaze that burned for hours. Outcry over the burning river led to the establishment of the Environmental Protection Agency (EPA) in 1970, whose mission is to protect the nation's air, water, and land. One of the EPA's first programs was the Clean Water Act (CWA), passed by Congress in 1972, which dictated that America's lakes, rivers, and estuaries must be free of pollution discharges by 1985. Although the Act's standard of 100 percent swimmable, fishable, and drinkable waters has yet to be reached, the CWA has forced industries and sewage plants to treat their wastewater effluent more effectively before discharging it into America's waterways. The EPA estimates that 64 percent of the rivers, 61 percent of the lakes, and 62 percent of the estuaries in the United States met the standards of the Clean Water Act in 1996.

When the EPA determines water quality standards and acceptable pollution levels for a specific river, lake, or other waterway, it considers the entire watershed and all the sources of water pollution in the watershed. The EPA then sets a total maximum daily load (TMDL) for certain pollutants that can enter that watershed without degrading the water quality. Getting industries and wastewater plants to treat their effluent before discharging it into the waterways is the easiest way to clean America's water. Such pollution is called point source pollution, because it enters the water from an easily defined source—usually drainage or sewage pipes or ditches. Nonpoint source pollution—which includes agricultural and urban runoff—is much more difficult to reduce and treat because the source of this type of water pollution cannot be clearly identified.

Agricultural runoff includes all the animal wastes and toxic chemicals used on fields. Farmers routinely apply manure, fertilizers, insecticides, and herbicides to their fields to boost crop yields. However, rainwater or irrigation may wash the soil and residues from these treatments into ditches, streams, lakes, and wetlands, thereby polluting the water supply. Rain also washes animal wastes into waterways, and some farmers use holding ponds, which sometimes leak, to store their animal wastes. The EPA is trying to make farmers more aware of these sources of water pollution and to take steps to reduce or prevent such pollution. Farmers are encouraged to stop plowing the remnants of their crops into their fields after harvesting to keep soil disturbances to a minimum, thus reducing water runoff. In addition, farmers are required to follow federal guidelines when storing or disposing of animal wastes.

A new form of pollution control hopes to reduce a watershed's pollution levels by targeting farm runoff. Effluent trading allows a polluter—such as a point source polluter—in one area of a watershed to increase its allowable water pollution levels if another polluter—such as a nonpoint source polluter—in the same watershed decreases its pollution discharge by at least the same amount. For example, the Rahr Malting Company, located on the banks of the Minnesota River in Shakopee, Minnesota, wanted to increase its production 20 percent, and under EPA rules it needed to build a wastewater treatment plant to handle its increased effluent. In order to receive a permit from the EPA for the treatment plant, Rahr agreed to reduce nonpoint source pollutants that entered the river upstream of the plant. To accomplish this, the company purchased easements along the river and reduced erosion of agricultural lands through stabilization and restoration of stream banks. These steps reduced upstream pollution enough to allow Rahr to increase its production almost 15 percent; the company hoped to achieve the remaining 5 percent in 1998.

Supporters of effluent trading contend that companies are allowed to reduce pollution in the most cost-effective manner. For example, a $1 million project that established buffer strips around waterways in the Tar-Pamlico River Basin in North Carolina and collected agricultural and animal waste runoff reduced nitrogen pollutants in the watershed by the same amount that a $7 million point source pollution control system would have. "Making pollution control more affordable may also speed the process by which water quality is improved and make it possible to clean up a greater volume of water," asserts Ric Jensen of the Texas Water Resources Institute. Moreover, he maintains, effluent trading encourages the public, industries, landowners, states, and local communities to work together to develop effective pollution control strategies.

Others are not so enthusiastic about the effluent trading program. According to the National Wildlife Federation (NWF), the EPA's effluent trading program merely trades one pollution site for another and does nothing to eliminate pollution discharges as mandated by the Clean Water Act. Moreover, shifting the site of the pollution discharge could have an enormous adverse impact on the waterway and environment, the federation maintains. For example, most point sources of water pollution—such as factories and treatment plants—have pollution discharge limits that can be monitored and enforced. However, most nonpoint sources of water pollution—farms, construction sites, logging areas, and suburban lawns—are unregulated and not easily monitored. Therefore, the NWF contends, it is difficult, if not impossible, to determine if measures to control nonpoint pollution runoff actually reduce pollution. Allowing polluters to trade pollution discharges may actually increase the amount of pollution in a watershed, the organization argues. "With approximately 40 percent of our waters still not meeting water quality goals, we cannot afford to move backwards," asserts Mark Van Putten, president of the NWF. If effluent trading becomes an accepted pollution control measure, Van Putten maintains that increased monitoring of a waterway's overall pollution levels will be necessary to ensure that pollution is reduced in actuality and not just on paper.

Water quality is vitally important to the health and safety of all the world's inhabitants. The debate over whether pollution credits such as effluent trading are an effective means of pollution control is just one issue that is examined in *Pollution: Opposing Viewpoints*. This anthology addresses several key questions about pollution in the following chapters: Is Pollution a Serious Problem? Do Chemical Pollutants Pose a Health Risk? Is Recycling an Effective Response to Pollution? How Can Air Pollution Be Reduced? How Should Pollution Be Managed? The authors examine the seriousness of pollution and how it can be reduced and managed.

CHAPTER 1

Is Pollution a Serious Problem?

CHAPTER PREFACE

Ever since Rachel Carson came out with her book *Silent Spring* in 1962 warning of the dangers of using DDT, environmentalists have been warning of the dangers of polluting the Earth. While many countries have seriously attempted to reduce their garbage and air and water pollution, others seem to be more concerned with increasing their economic output.

Studies show that in the developed, western nations, many of the world's pollution problems are declining. A 1996 report by the Organization for Economic Cooperation and Development found that the emission of gases such as chlorofluorocarbons (CFCs) that are believed to cause global warming have dropped by two-thirds since the 1980s. Improvements in wastewater treatments have also led to a reduction of raw sewage that is dumped into the developed countries' waterways. The amount of contaminated land and water in First World countries is also falling as their governments tighten the regulations concerning the disposal of toxic wastes.

Environmentalists point out that in the Third World countries, however, pollution is still a very serious problem. Developing nations are often more concerned about improving their economies, they argue, than with spending money to protect their environment. For example, inadequate or nonexistent sewage treatment facilities pollute the water supply, and millions of people are killed annually by drinking or using contaminated water, and some industrial areas of China have the most polluted air in the world. These environmentalists assert that it is only when the poorer countries reach a level of economic prosperity that they begin to think about taking care of the environment.

A global overview shows that while pollution problems may be improving in one part of the world, they may be worsening in another. The authors in the following chapter examine whether garbage, air, and water pollution are a serious problem.

VIEWPOINT 1

"The United States is sinking under a 'river of waste.'"

GARBAGE POLLUTION IS A SERIOUS PROBLEM

Lynn Landes

Lynn Landes is the founder and director of Zero Waste America, an environmental organization dedicated to the elimination of waste and pollution through legislative reform and changing practices. In the following viewpoint, Landes argues that the United States has a waste disposal problem that is out of control. She maintains that the country needs a national plan to reduce, eliminate, or recycle waste, and that states should be allowed to refuse the importation of domestic and international solid waste.

As you read, consider the following questions:

1. According to Zero Waste America, how much foreign and domestic waste was disposed of per person in the United States in 1997?
2. What is Pennsylvania's claim to fame, according to Landes?
3. What three courses of action does Landes recommend to eliminate waste and waste imports?

Reprinted from Lynn Landes, "River of Waste," 1997 web article at www.zerowasteamerica.com/WasteWatch.htm, by permission of the author.

The United States is sinking under a "river of waste." Zero Waste America (ZWA) estimates that in 1997, Americans will dispose of more than 1.2 billion tons of domestic and imported waste. That amounts to approximately 5 tons of waste disposed for every person in the country. The cost to public health and natural resources is incalculable.

In 1970, Congress passed legislation establishing National Environmental Policy to "enhance the quality of renewable resources and approach the maximum attainable recycling of depletable resources." That policy was a mandate from Congress. It should serve as a lifeboat to a sustainable environment, to no more landfills or incinerators, and to Zero Waste.

NO NATIONAL PLAN TO REDUCE WASTE

Instead, the United States has no effective national plan to eliminate or reduce waste. There is no government effort to create sustainable markets for recyclables. There is no national ban on the disposal of waste, compost, or recyclables into landfills or incinerators. There is no limit on the amount of waste imported from other countries. The EPA does not even track the total amount of waste that is generated, imported, or disposed in the United States.

The free market has not provided a foundation this nation needs to reduce, eliminate, or recycle waste. Voluntary programs of waste recycling and reduction have not been sufficient to curb the ever-increasing need to build more landfills and incinerators.

FAILED POLICIES AND ENFORCEMENT

Bucks County, Pennsylvania, is a good example of failed policies and enforcement by both federal and state authorities. Bucks County disposes of approximately 2,000 tons of county waste daily. Waste Management (WMX) is permitted by Pennsylvania's Department of Environmental Protection (DEP) to dispose of 20,000 tons of waste each day in Bucks County. Federal law requires that states must have a "state solid waste management plan" to ensure maximum recycling and resource conservation, and to assess environmental impact of waste disposal facilities. Pennsylvania has no such state plan, yet the Environmental Protection Agency (EPA) allows the state to continue accepting waste and issuing permits for more disposal facilities. Currently, Pennsylvania is the leading importer of foreign and domestic waste in the nation.

Many states complain that waste imports undercut their efforts at waste reduction and recycling. For the last several years, states have looked to proposed federal legislation that promises states protection from imports. This proposed legislation will not protect states for three reasons:

• The proposed legislation only applies to "unwanted" waste. A state cannot prohibit a municipality from accepting waste, if an agreement is reached between the host municipality and a waste disposal company. This invites the waste industry to "shop" for disadvantaged communities who may want the host fees to offset tax increases, or can't afford to defend themselves against well-funded waste industry legal action.

• There is no limit on other types of disposal waste that can be imported from other states or nations. Only "municipal" waste will be affected by this legislation. That may account for as little as 20% of all waste disposed in a state.

• This legislation will encourage the importation of more toxic waste, such as: hazardous, industrial, infectious, asbestos, sewage sludge, contaminated soil, and incinerator ash. Much of this waste is allowed in municipal landfills, as well as in private and commercial landfills and incinerators. [The Legislation has not been passed into law.]

A ZERO WASTE PLAN FOR THE FUTURE

So, what's the answer? In the absence of Congressional action or federal enforcement of current environmental law, how do states eliminate waste and protect themselves from waste imports?

First, states can issue waste "disposal bans" for both in-state and imported waste. They can begin by banning compostables,

Jerry Barnett/The Indianapolis News. Reprinted with permission.

such as food and yard waste. A general rule is that waste must be free from hazardous materials in order to be composted or recycled safely.

Second, states can legislate a variety of measures to sustain recycling markets. They can set minimum recycled content standards and establish bottle bills and other "take-back" legislation. With markets guaranteed, recyclables can be banned for disposal.

Lastly, states should store hazardous waste until it can be safely recycled. Never bury or burn waste!

In order to withstand legal challenge by waste importers, states must apply disposal bans equally to both in-state and out-of-state waste. In the *City of Philadelphia v. New Jersey* (1978), the Supreme Court ruled that New Jersey could protect its environment in the following statement, *"And it may be assumed as well that New Jersey may pursue those ends by slowing the flow of all waste into the State's remaining landfills, even though interstate commerce may incidentally be affected."* Again, in *National Solid Waste Association v. Meyer* (1995), Federal Court of Appeals, 7th Circuit ruled, *"Accordingly, Wisconsin could realize its goals of conserving landfill space and protecting the environment by mandating that all waste entering the State first be treated at a materials recovery facility with the capacity to effect this separation."* Note that the words "all waste" were used in both decisions.

As a nation, we can turn this "river of waste" into a "reservoir of recyclables." We should do whatever it takes to eliminate waste. Zero Waste is our goal. A healthy and clean environment . . . let that be our legacy.

"If Americans were really creating more trash by overindulging, we would be spending more on trash-generating items. . . . But household expenditures for nondurable goods . . . declined."

GARBAGE POLLUTION HAS IMPROVED

Robert M. Lilienfeld and William L. Rathje

Robert M. Lilienfeld and William L. Rathje are publishers of the *ULS Report* (Use Less Stuff), a newsletter about preventing waste. In the following viewpoint, the authors argue that although the amount of garbage and waste being generated must be reduced, the problem of waste disposal is not as bad as many people believe. In fact, some aspects of waste disposal have even improved over the years. The best ways to continue to reduce the amount of garbage, Lilienfeld and Rathje maintain, is to improve products so they do not need to be replaced as frequently and to reduce and reuse the packaging for the products.

As you read, consider the following questions:

1. What is the real problem with landfills, according to the authors?
2. How do Lilienfeld and Rathje respond to the claim that packaging is responsible for the garbage problem?
3. In the authors' opinion, why does the amount of garbage generated per American continue to increase?

We participated in an environmental festival at the Mall of America in Bloomington, Minn., the largest indoor shopping center in the country. Having spoken with literally thousands of parents, children and teachers, we were appalled at the public's wealth of environmental misunderstanding.

We were equally chagrined by the superficiality of what we heard, and have coined a new term for this type of sound-bite-based, factoid-heavy understanding: eco-glibberish. Here are half a dozen examples:

RECYCLING IS THE KEY

Myth: *The most important thing we can do is to recycle.* Actually, it's one of the least important things we can do, if our real objective is to conserve resources. Remember the phrase "reduce, reuse and recycle"? Reduce comes first for a good reason: it's better to not create waste than to have to figure out what to do with it. And recycling, like any other form of manufacturing, uses energy and other resources while creating pollution and greenhouse gases.

Rather, we need to make products more durable, lighter, more energy efficient and easier to repair rather than to replace. Finally, we need to reduce and reuse packaging.

GARBAGE WILL OVERWHELM US

Myth: *There's a garbage crisis.* The original garbage crisis occurred when people first settled down to farm and could no longer leave their campsites after their garbage grew too deep.

Since then, every society has had to figure out what to do with discards. That something was usually unhealthy, odiferous and ugly—throwing garbage in the streets, piling it up just outside of town, incorporating it into structures or simply setting it on fire.

Today we can design history's and the world's safest recycling facilities, landfills and incinerators. We even have a national glut of landfill capacity, thanks to the fact that we've been building large regional landfills to replace older, smaller local dumps.

The problem is political. No one wants to spend money on just getting rid of garbage or to have a garbage site in the backyard.

The obvious solution is to stop generating so much garbage in the first place. Doing so requires both the knowledge and self-discipline to conserve energy and to do more with less stuff.

INDUSTRY IS TO BLAME

Myth: *It's all industry's fault.* No, it's all people's fault. Certainly industry has played a significant role in destroying habitats, generating pollution and depleting resources. But we're the ones who signal businesses that what they're doing is acceptable—every time we open our wallets.

And don't just blame industrial societies. In his recent book *Earth Politics*, Ernst Ulrich von Weizsäcker wrote that "perhaps 90 percent of the extinction of species, soil erosion, forest and wilderness destruction and also desertification are taking place in developing countries." Thus, even non-industrialized, subsistence economies are creating environmental havoc.

THE EARTH IS IN PERIL

Myth: *We have to save the earth.* Frankly, the earth doesn't need to be saved.

Nature doesn't give a hoot if human beings are here or not. The planet has survived cataclysmic and catastrophic changes for millions upon millions of years. Over that time, it is widely believed, 99 percent of all species have come and gone while the planet has remained.

Saving the environment is really about saving *our* environment—making it safe for ourselves, our children and the world as we know it. If more people saw the issue as one of saving themselves, we would probably see increased motivation and commitment to actually doing so.

Myth: *Packaging accounts for a growing percentage of our solid waste.* If you were to examine a dumpster of garbage from the 1950's and a dumpster of garbage from the 1980's, you would find more discarded packaging in the first one. Packaging has actually decreased as a proportion of all solid wastes—from more than half in the 1950's to just over one-third today.

One reason is that there was more of other kinds of wastes—old appliances, magazines, office paper—in the 1980's. But the main causes were two changes in the packaging industry.

First, the heavy metal cans and glass bottles of the 1950's gave way to far lighter and more crushable containers—about 22 percent lighter by the 1980's. At the same time, many metal and glass containers were replaced by paper boxes and plastic bottles and bins, which are even lighter and more crushable.

Second, the carrying capacity of packages—the quantity of product that can be delivered per ounce of packaging material—increased hugely.

Glass, for example, has a carrying capacity of 1.2, meaning that 1.2 fluid ounces of milk or juice are delivered for every ounce of glass in which they are contained. Plastic containers have a carrying capacity of about 30.

AMERICANS ARE WASTING MORE

Myth: *Americans are over-consumers, since the per capita creation of solid waste continues to climb.* Each person generates about 4.4 pounds of garbage a day—a number that has been growing steadily. The implication is that we partake in an unstoppable orgy of consumption. The truth is far more mundane.

In reality, increases in solid waste are based largely on the mathematics of households, not individuals. That's because regardless of the size of a household, fixed activities and purchases generate trash.

As new households form, they create additional garbage. Think about a couple going through a divorce. Once there was one home. Now there are two. Building that second house or condo used lots of resources and created lots of construction debris. Where once there was one set of furniture, one washing machine and one refrigerator, now there are two. Each refrigerator contains milk cartons, meat wrappers and packages of mixed vegetables. Each pantry contains cereal boxes and canned goods.

THE AMOUNT OF GARBAGE IS DECREASING

According to figures from the American Public Works Association and the Environmental Protection Agency, as late as 1939 cities like Newark, New Jersey, and Austin, Texas, reported annual per capita discards of garbage, ash, and rubbish 20 percent greater than the refuse of the average American in 1988. True, affluence may cause more discards (since in one sense affluence *means* there are more things per capita to eventually discard). But along with a higher standard of living come phenomena such as "light-weighting," where the producers of a commodity (e.g., plastic cola bottles and aluminum cans) find a way to produce the same service with less material. As Judd Alexander observes, fast-food restaurants also help decrease waste: a typical McDonald's discards less than two ounces of garbage for each customer served.

Richard Shedenhelm, *Freeman*, April 1995.

To make matters worse, households are growing at a fairly rapid rate, almost double the rate of population growth. That's because we're all living longer and away from our children, divorcing more frequently and becoming far more accepting of single-parent households.

Census Bureau numbers tell this story: From 1972 to 1987, the population grew by 16 percent. The number of households grew by 35 percent. Municipal solid waste increased by 35 percent, too.

If Americans were really creating more trash by overindulging, we would be spending more on trash-generating items: nondurable goods like food and cosmetics. These all generate lots of garbage, since they are used and discarded quickly, along with their packaging. But household expenditures for nondurable goods, as measured by constant dollars, declined slightly from 1972 to 1987—by about one-half of 1 percent.

Does all of this mean we can sit back and relax? No. The earth's resources are finite. Habitats are being destroyed. Biodiversity is declining. And the consumption of resources is expanding.

But it does mean that we must be less willing to accept glib, ideological pronouncements of right and wrong, good and evil, cause and effect. Thus, to truly change the world for the better, we need more facts, not simply more faith.

> "Air pollution from the combustion
> of fossil fuels (oil, coal, and natural
> gas) in cars, trucks, and power
> plants, is killing roughly 60,000
> Americans each year."

AIR POLLUTION IS A SERIOUS PROBLEM

Peter Montague

In the following viewpoint, Peter Montague argues that air pollution is responsible for the health problems and even deaths of hundreds of thousands of Americans. The particles that make up this hazardous air pollution are so small that they can easily get into the deepest part of the lungs where they can cause inflammation and other damage, he asserts. Although numerous studies have shown the harmful effects of air pollution on human health, Americans remain unprotected against the dangers of air pollution due to unreasonable demands by U.S. regulatory agencies, he contends. Montague is the director of the Environmental Research Foundation and editor of *Rachel's Hazardous Waste News*.

As you read, consider the following questions:

1. According to a study by researchers at Harvard University School of Public Health, how many Americans die each year from the effects of air pollution?
2. According to Montague, what is the danger to humans from PM2.5?
3. In the author's opinion, what is preventing U.S. regulatory agencies from regulating PM2.5?

Reprinted from Peter Montague, "The Holy Grail of Scientific Certainty," *Rachel's Environment and Health Weekly*, May 4, 1995, by permission of the Environmental Research Foundation. Endnotes in the original publication have been omitted in this reprint.

Air pollution from the combustion of fossil fuels (oil, coal, and natural gas) in cars, trucks, and power plants, is killing roughly 60,000 Americans each year, according to researchers at Harvard University's School of Public Health. This represents about 3% of all U.S. deaths each year. Every combustion source is contributing to the death toll; none is benign, including incinerators; soil burners; flares and after-burners; industrial and residential heaters and boilers; cars; buses; trucks; and power plants. Diesel vehicles and oil- and coal-burning power plants seem to be the worst offenders.

FINE-PARTICLE POLLUTION

The culprit in every case is the fine particles—invisible soot—created by combustion. Fine particles are not captured efficiently by modern pollution-control equipment. Furthermore, they are not visible except as a general haze. They are far too small to be seen.

According to more than a dozen studies, there seems to be no threshold, no level of fine-particle pollution below which no deaths occur. Even air pollution levels that are well within legal limits are killing people, especially older people, and people with chronic heart and lung ailments, the Harvard researchers have found.

Furthermore, studies indicate that fine-particle pollution is causing or exacerbating a wide range of human health problems, including: initiating, and worsening, asthma, especially in children; increasing hospital admissions for bronchitis, asthma, and other respiratory diseases; increasing emergency room visits for respiratory diseases; reducing lung function (though modestly) in healthy people as well as (more seriously) in those with chronic diseases; increasing upper respiratory symptoms (runny or stuffy nose; sinusitis; sore throat; wet cough; head colds; hay fever; and burning or red eyes); and increasing lower respiratory symptoms (wheezing; dry cough; phlegm; shortness of breath; and chest discomfort or pain); and heart disease.

Since 1987, U.S. Environmental Protection Agency (EPA) has been measuring fine-particle air pollution, calling it PM10, which means "particulate matter 10 micrometers or less in diameter." A micrometer is a millionth of a meter, and a meter is about a yard. The dot above the letter i in a typical newspaper measures about 400 micrometers in diameter.

EPA measures PM10 pollution by weight—the total weight of all particles with a diameter of 10 micrometers or less in each cubic meter of air. The legal limit is 50 micrograms of PM10 particles in each cubic meter of air, as a year-round average. (A gram is ⅟₂₈th of an ounce, and a microgram is a millionth of a gram.) Many U.S., Canadian, and European cities from time to time will have as much as 100 to 200 micrograms of PM10 particles in each cubic meter of air.

SIZE MATTERS

The size of the particles is what's most important from a public health viewpoint. Particles larger than 10 micrometers generally get caught in your nose and throat, never entering the lungs. Particles smaller than 10 micrometers can get into the large upper branches just below your throat where they are caught and removed (by coughing and spitting or by swallowing). Particles smaller than 5 micrometers can get into your bronchial tubes, at the top of the lungs; particles smaller than 2.5 micrometers in diameter can get down into the deepest (alveolar) portions of your lungs where gas exchange occurs between the air and your blood stream, oxygen moving in and carbon dioxide moving out. These are the really dangerous particles because the deepest (alveolar) portions of the lung have no efficient mechanisms for removing them. If these particles are soluble in water, they pass directly into the blood stream within minutes. If they are not soluble in water, they are retained in the deep lung for long periods (months or years).

About 60% of PM10 particles (by weight) have a diameter of 2.5 micrometers or less. These are the particles that can enter the human lung directly. (They also enter homes; indoor air and outdoor air typically contain the same quantities of fine particles, so buildings provide no refuge from these invisible killers.) Let's go back to the dot over the letter i. If particles are 10 micrometers in diameter, then 1,600 particles can fit on the dot. If the particle diameter is 2.5 micrometers, then 25,600 particles can fit on the dot. When the diameter drops to 0.3 micrometers, then 1.8 million particles can fit on the dot, and when the diameter is reduced to 0.001 micrometers, or one nanometer, then 160 billion (1.6^{11}) particles can fit on the dot over the letter i.

In a modern U.S. city, on many days, the air will contain 100 billion (1^{11}) one-nanometer-diameter particles in each cubic meter of air, all of them invisible. By weight, these 100 billion particles will only amount to 0.00005 micrograms (one ten-thousandth of 1 percent of the 50-microgram legal limit), yet they may be responsible for much of the health damage created by fine-particle pollution. For this reason, in 1979, the National Research Council said that measuring particles by weight, without regard to particle size, has "little utility for judging effects." Particle size is everything when it comes to air pollution and health.

A CLEAR RELATIONSHIP

The study of fine particles and their effects on human health has been under way in earnest since 1975. Since then, studies have been able to rule out sulphur dioxide and ozone pollution as the cause of the observed deaths. In 1995, a new study of 552,138 adult Americans in 151 metropolitan areas confirmed once again that there is a clear relationship between fine-particle air pollution and human deaths, and it ruled out smoking as a cause of the observed deaths. This study is particularly important because it didn't simply match death certificates with pollution levels; it actually examined the characteristics (race, gender, weight and height) and lifestyle habits of all 552,138 people. Thus the study was able to rule out tobacco smoking (cigarettes, pipe and cigar); exposure to passive tobacco smoke; occupational exposure to fine particles; body mass index (relating a person's weight and height); and alcohol use. The new study also controlled for changes in outdoor temperature. The study found that fine-particle pollution was related to a 15% to 17% difference in death rates between the least-polluted cities and the most-polluted cities.

Up until this year, researchers have shown that fine particles cause death and disease, but the mechanism by which this occurs has remained a mystery. A new hypothesis, published in January 1995 in the British medical journal Lancet, suggests that the particles retained in the deep lung cause inflammation which, in turn, releases natural chemicals into the blood stream causing coagulation of the blood. This hypothesis is proposed as the biological mechanism by which fine particles cause respiratory and heart-related diseases and death.

One might ask why steps weren't taken long ago to prevent the disease and death associated with fine-particle pollution. After all, the National Academy of Sciences said as early as 1979 that ". . . alveolar retention of relatively insoluble particles is recognized to be important to the pathogenesis of chronic lung disease. . . ." In other words, the Academy was convinced in 1979 that fine particles retained in the deep lung cause long-term lung disease. The Academy said at that time, "In summary, particulate atmospheric pollutants may be involved in chronic lung disease pathogenesis as causal factors in chronic bronchitis, as predisposing factors to acute bacterial and viral bronchitis, especially in children and cigarette smokers, and as aggravating factors for acute bronchial asthma and the terminal stages of oxygen deficiency (hypoxia) associated with chronic bronchitis and/or emphysema and its characteristic form of heart failure (cor pulmonale)."

Clay Bennett/North America Syndicate. Reprinted with permission.

If the Harvard researchers are correct in their estimate, that 60,000 Americans die each year from fine-particle pollution, and tens of thousands more are made sick (especially children), then we can calculate that, since 1979, nearly a million Americans (960,000) have been killed by fine-particle pollution, and millions more have been made sick.

PROVING HARM

Why can't we act to prevent this important problem? Because U.S. regulatory agencies—and the courts—have lost their way, searching for the holy grail of scientific certainty. Regulators and judges now insist that science has to "prove harm" before regulatory control can begin. Philosophers of science know that science cannot "prove" anything. It often takes science decades—sometimes centuries—to reach a clear majority opinion and there will always be uncertainties, giving rise to nagging doubts, which can only be laid to rest by further study. In the meantime, the science of 1979 was sufficient to tell us that people are dying and children are getting sick because of fine particles. The precise mechanism of harm is, even today, not fully understood, but the harm itself has been clear beyond any reasonable doubt for many years. Common sense says that the National Academy's conclusions back in 1979 should have been sufficient for regulators to clamp down in earnest.

The long history of fine-particle research raises a series of difficult questions: When did our society first turn away from the common sense, weight-of-the-evidence, preventive approach which, theoretically at least, guides public health decisions? What role did corporate lawyers and scientists play in convincing scientifically-illiterate judges and politicians that scientific certainty was required before society could take prudent steps to protect public health and safety? What steps can concerned citizens take NOW to move our society back toward a prevention-based approach to the control of dangerous materials and technologies?

"All the years throughout the 1990s
have had better air quality than any
of the years in the 1980s, showing
a steady trend of improvement."

AIR POLLUTION IS DECLINING

Environmental Protection Agency

The Environmental Protection Agency (EPA) is the federal
agency in charge of protecting the environment and controlling
pollution. The following viewpoint is a summary of the EPA's
National Air Quality and Emissions Trends Report of 1996. According to
the EPA's report, emission levels for most air pollutants declined
during the 1990s. These levels of air pollutants decreased even
though the U.S. population, economy, and the number of vehicle miles traveled all increased. The EPA has reduced some air
pollutant standards even further, which the organization claims
will lead to lower levels of pollution.

As you read, consider the following questions:

1. What is the only air pollutant whose emission level has
   increased since 1970, according to the EPA?
2. What would account for a difference in the percent change in
   ambient concentrations of an air pollutant and the percent
   change in emissions, in the author's opinion?
3. What has historically been the largest source of lead
   emissions, according to the EPA?

Reprinted from "National Air Quality and Emissions Trends Report, 1996," of the Office
of Air Quality Planning and Standards, U.S. Environmental Protection Agency.

S ince 1970, air quality has continued to improve for six major air pollutants known as "criteria" pollutants. They include carbon monoxide, lead, nitrogen dioxide, ground-level ozone, particulate matter, and sulfur dioxide. Emissions of all of these pollutants except nitrogen oxides have decreased significantly since 1970. Between 1970 and 1996 emissions of NOx have increased 8 percent. Emissions of NOx contribute to the formation of ozone. In October 1997, the Environmental Protection Agency (EPA) proposed a rule that will significantly reduce emissions of NOx in 22 eastern states, and, in turn, reduce the regional transport of ozone.

The improvements in air quality and economic prosperity that have occurred since EPA initiated air pollution control programs in the early 1970s illustrate that economic growth and environmental protection can go hand-in-hand. Since 1970, national total emissions of the six criteria pollutants declined 32 percent, while U.S. population increased 29 percent, gross domestic product increased 104 percent, and vehicle miles traveled increased 121 percent.

Nationally, the 1996 air quality levels are the best on record for all six criteria pollutants. In fact, all the years throughout the 1990s have had better air quality than any of the years in the 1980s, showing a steady trend of improvement.

Despite continued improvements in air quality, approximately 46 million people lived in counties that did not meet the air quality standards for at least one of the six criteria pollutants in 1996.

AIR QUALITY TRENDS

For the past 24 years, EPA has examined the air pollution trends of the six criteria pollutants. The 1996 National Air Quality and Emissions Trends Report discusses the nation's progress in cleaning up these pollutants during the 10-year period between 1987 and 1996.

In the report, EPA tracks trends in concentrations and emissions of these six pollutants. Air quality information is based on actual measurements of pollutant concentrations in the air from nearly 5,000 monitoring sites located throughout the nation. Air pollutant emission trends are based on engineering estimates of the total tonnage these pollutants released to the air annually. However, starting in 1994, under the Acid Rain Program, EPA began tracking measured emission values of sulfur dioxide and nitrogen oxides based on actual emissions data from continuous emission monitors for the electric utility industry.

Generally there are similarities between air quality trends and emission trends for any given pollutant. However, in some cases, there are notable differences between the percent change in ambient concentrations and the percent change in emissions. These differences can mainly be attributed to the location of air quality monitors. Most monitors are positioned in urban, population-oriented locales which are more likely to indicate reductions in emissions that occur in urban areas (such as emissions from automobiles) rather than emissions that occur in rural areas (such as emissions from power plants). Thus, trends in air quality more closely track changes in urban emissions rather than changes in total national emissions.

In July 1997, EPA revised the ozone and particulate matter standards following a lengthy scientific review process. Prior to this time, the PM standard applied to particles whose aerodynamic size is less than or equal to 10 micrometers, or PM_{10}. The National Ambient Air Quality Standards (NAAQS) revision strengthened protection against particles in the smaller part of that range by adding an indicator for $PM_{2.5}$ (those whose aerodynamic size is less than or equal to 2.5 micrometers). The combination of the PM_{10} and $PM_{2.5}$ indicators will provide protection against a wide array of particles.

CARBON MONOXIDE (CO)

Carbon monoxide enters the bloodstream through the lungs and reduces oxygen delivery to the body's organs and tissues. The health threat from CO is most serious for those who suffer from cardiovascular disease. At higher levels of exposure, healthy individuals are also affected. Visual impairment, reduced work capacity, reduced manual dexterity, poor learning ability, and difficulty in performing complex tasks are all associated with exposure to elevated CO levels. Automobiles are a large source of emissions of carbon monoxide.

Over the past 10 years, ambient concentrations of CO decreased 37 percent, and the estimated number of exceedances of the 8-hour standard decreased 92 percent. Also, CO emissions decreased 18 percent, and CO emissions from highway vehicles decreased 26 percent. These improvements occurred despite a 28 percent increase in vehicle miles traveled during this 10-year period. Between 1995 and 1996, ambient CO concentrations decreased 7 percent and emissions of CO decreased 1 percent.

LEAD (PB)

Exposure to lead occurs mainly through the inhalation of air and the ingestion of lead in food, water, soil, or dust. It accumulates in the blood, bones, and soft tissues. Because it is not readily excreted, lead can also adversely affect the kidneys, liver, nervous system, and other organs. Excessive exposure to lead may cause neurological impairments such as seizures, mental retardation, and/or behavioral disorders. At low doses, fetuses and children often suffer from central nervous system damage. Recent studies also show that lead may be a factor in high blood pressure and subsequent heart disease.

Historically, automobiles have been the largest source of emissions of lead. However, the widespread use of unleaded gasoline has dramatically reduced the contribution from automobiles worldwide. Industrial processes (e.g., metals processing) are another significant source of lead emissions.

Between 1987 and 1996, ambient lead concentrations decreased 75 percent, and lead emissions decreased 50 percent. Lead emissions from highway vehicles have decreased 99 percent since 1987 as a result of the increased use of unleaded gasoline and the reduction of the lead content in leaded gasoline. Between 1995 and 1996, lead concentrations remained unchanged, total lead emissions decreased 2 percent, and lead emissions from transportation sources did not change. While lead emissions from industrial sources have dropped more than 90 percent since the late 1970s, some serious point-source lead problems remain.

NITROGEN DIOXIDE (NO_2)

Nitrogen dioxide can irritate the lungs and lower resistance to respiratory infections such as influenza. The effects of short-term exposure are still unclear, but continued or frequent exposure to concentrations higher than those normally found in the ambient air may cause increased incidence of acute respiratory disease in children. Nitrogen oxides are an important precursor to both ozone and acidic precipitation (acid rain) and can affect both terrestrial and aquatic ecosystems. The regional transport and deposition of nitrogenous compounds arising from emissions of NOx is a potentially significant contributor to such environmental effects as the growth of algae and subsequent unhealthy or toxic conditions for fish in the Chesapeake Bay and other estuaries. In some parts of the western United States, NOx have a significant impact on particulate matter concentrations. Automobiles and fuel combustion (e.g., electric utilities) are a significant source of emissions of nitrogen dioxide.

Between 1987 and 1996, ambient concentrations of NO_2 decreased 10 percent, but total emissions of nitrogen oxides (NOx) increased 3 percent, due primarily to increased emissions from non-utility fuel combustion. Between 1995 and 1996, national average annual mean NO_2 ambient concentrations remained unchanged, while total emissions of NOx decreased 2 percent. Emissions from highway vehicles, also a source of NOx emissions, decreased 6 percent between 1987 and 1996, while NOx emissions from utility fuel combustion decreased 3 percent.

Ozone (O_3)

Ozone occurs naturally in the stratosphere and provides a protective layer high above the earth. At ground-level, however, it is the prime ingredient of smog. Short-term exposures (1 to 3 hours) to ambient ozone concentrations have been linked to increased hospital admissions and emergency room visits for respiratory causes. Repeated exposures to ozone can make people more susceptible to respiratory infection and lung inflammation, and can aggravate preexisting respiratory diseases such as asthma. Other health effects attributed to short-term exposures to ozone, generally while individuals are engaged in moderate or heavy exertion, include significant decreases in lung function and increased respiratory symptoms such as chest pain and cough.

Children active outdoors during the summer when ozone levels are at their highest are most at risk of experiencing such effects. Other at-risk groups include outdoor workers, individuals with preexisting respiratory disease such as asthma and chronic obstructive lung disease, and individuals who are unusually responsive to ozone. Recent studies have attributed these same health effects to prolonged exposures (6 to 8 hours) to relatively low ozone levels during periods of moderate exertion. In addition, long-term exposures to ozone present the possibility of irreversible changes in the lungs which could lead to premature aging of the lungs and/or chronic respiratory illnesses.

The regional transport of ozone is a problem, particularly in the eastern United States. Ozone is not emitted directly into the air, rather it is formed when volatile organic compounds (VOC) react in the presence of sunlight. In order to address ozone pollution, EPA has traditionally focused its control strategies on reducing emissions of volatile organic compounds in individual nonattainment areas. However, EPA and the States have recognized a need for an aggressive program to reduce regional emissions of nitrogen oxides. In October 1997, EPA proposed a rule that will significantly reduce emissions of NOx in 22 eastern states, and, in turn, reduce the regional transport of ozone. National trends in emissions of NOx and VOC underscore the importance of this new approach. Between 1970 and 1996 emissions of VOCs have decreased 38 percent whereas emissions of NOx have increased 8 percent. Further, between 1987 and 1996 emissions of VOCs have decreased 18 percent whereas emissions of NOx have increased 3 percent.

Between 1987 and 1996, ambient ozone concentrations decreased 15 percent, and the estimated number of exceedances of the ozone standard decreased 73 percent. Between 1995 and 1996, ambient ozone concentrations decreased 6 percent, while the estimated number of exceedances of the ozone standard decreased 37 percent. VOC emissions decreased 18 percent between 1987 and 1996 and decreased 7 percent between 1995 and 1996. NOx emissions, the other main ozone precursor, increased 3 percent between 1987 and 1996 and decreased 2 percent from 1995 to 1996.

PERCENT CHANGE IN NATIONAL AIR QUALITY CONCENTRATIONS AND EMISSIONS

	Air Quality Concentration % Change 1977–1996	Emissions % Change 1970–1996	Air Quality Concentration % Change 1987–1996	Emissions % Change 1987–1996
Carbon Monoxide	−61%	−31%	−37%	−18%
Lead	−97%	−98%	−75%	−50%
Nitrogen Dioxide	−27%	+8% (NOx)	−10%	+3% (NOx)
Ozone	−30%	−38% (VOC)	−15%	−18% (VOC)
PM_{10}	Data N/A	−73%+	−25%*	−12%+*
Sulfur Dioxide	−58%	−39%	−37%	−14%

*Based on 1988 to 1996 data.

+Includes only directly emitted particles. Secondary PM formed from SOx, NOx, and other gases comprise a significant fraction of ambient PM.

Environmental Protection Agency, National Air Quality and Emissions Report, 1996, 1997.

On July 18, 1997, EPA revised the national ambient air quality standards for ozone and particulate matter. After a lengthy scientific review process, including extensive external scientific review, and public review and comment, EPA issued a rule that will replace the previous 1-hour ozone 0.12 parts per million (ppm) standard with a new 8-hour 0.08 ppm standard to better protect public health and the environment. Although areas that do not meet the new 8-hour standard will not be designated "nonattainment" until the year 2000, the National Air Quality and Emissions Trends Report can begin to track trends in 8-hour levels of ozone. Nationally, 8-hour levels of ozone have decreased 11 percent over the past 10 years.

Particulate Matter (PM_{10})

Scientific studies show a link between particulate matter (alone, or combined with other pollutants in the air) and a series of significant health effects. These health effects include premature death, increased hospital admissions and emergency room visits, increased respiratory symptoms and disease, decreased lung function, and alterations in lung tissue and structure and in respiratory tract defense mechanisms. Sensitive groups that appear to be at greatest risk to such effects include the elderly, individuals with cardiopulmonary disease such as asthma, and children. In addition to health problems, particulate matter is the major cause of reduced visibility in many parts of the United States. Airborne particles also can cause soiling and damage to materials. Emissions of PM_{10} come from a variety of sources including windblown dust and grinding operations.

Ambient PM_{10} concentrations decreased 25 percent between 1988 and 1996 and decreased 4 percent between 1995 and 1996. PM_{10} estimated emissions (excluding fugitive emissions and emissions from natural sources) decreased 12 percent between 1988 and 1996 and remained unchanged between 1995 and 1996.

On July 18, 1997, EPA revised the national ambient air quality standards for particulate matter. After a lengthy scientific review, EPA determined that the annual PM_{10} standard set at 50 micrograms per cubic meter ($\mu g/m^3$) should not change but that the form of the PM_{10} 24-hour standard (which remains at a level of 150 $\mu g/m^3$) should be revised. Further, these studies indicated serious health risk associated with exposure to particles in the smaller part of that range. Therefore, EPA added an indicator for $PM_{2.5}$ (those whose aerodynamic size is less than or equal to 2.5 micrometers). The combination of the PM_{10} and $PM_{2.5}$ indicators will provide protection against a wide array of particles. The annual $PM_{2.5}$ standard is set at 15 $\mu g/m^3$ and the 24-hour $PM_{2.5}$ standard set at 65 $\mu g/m^3$. The secondary (welfare-based) standards were also revised by making them identical to the primary (health-based) standards. In conjunction with the Regional Haze Program, the secondary standards will protect against major PM welfare effects, such as visibility impairment, soiling, and materials damage. In July, 1997, EPA proposed a rule to address regional haze, and EPA plans to finalize this rule in July, 1998.

As the implementation of these revised standards begins and

a national air quality monitoring network for $PM_{2.5}$ is established, the *National Air Quality and Emissions Trends Report* will track trends in concentrations of $PM_{2.5}$.

SULFUR DIOXIDE (SO_2)

The major health concerns associated with exposure to high concentrations of SO_2 include effects on breathing, respiratory illness, alterations in the lungs' defenses, and aggravation of existing cardiovascular disease. Major subgroups of the population that are most sensitive to SO_2 include asthmatics and individuals with cardiovascular disease or chronic lung disease, as well as children and the elderly. Fuel combustion (e.g., from electric utilities) is a significant source of emissions of sulfur dioxide.

Between 1987 and 1996, ambient concentrations of SO_2 decreased 37 percent, while emissions of SO_2 decreased 14 percent. Between 1995 and 1996, nationwide average ambient SO_2 concentrations remained unchanged, while SO_2 emissions increased 3 percent. SO_2 emissions from electric utilities decreased 20 percent between 1987 and 1996. Between 1995 and 1996, SO_2 emissions from electric utilities increased 4 percent. The recent reductions in SO_2 emissions from electric utilities (down 17 percent since 1993) are due, in large part, to controls implemented under EPA's Acid Rain Program. The increase in SO_2 emissions that occurred between 1995 and 1996 is primarily due to increased demand for electricity.

"With frightening rapidity, . . . the
[oceans are] being devastated by
human activities . . . pollution from
industrial and household wastes and
agricultural run-off [and] massive
fish die-offs caused by oil spills."

WATER POLLUTION IS A SERIOUS PROBLEM

Dick Russell

In the following viewpoint, Dick Russell argues that the marine
life along many of the Earth's coastlines and in its oceans have
been killed by pollution from sewage effluent and fertilizer and
pesticide run-offs. Although some of these areas have made a remarkable comeback once pollutants are no longer dumped in
them, he asserts that cleaning up the oceans will take a coordinated worldwide effort. Russell is a freelance writer living in
Boston who has written several articles about the oceans.

As you read, consider the following questions:

1. What percentage of the Earth's population lives 50 miles or
   less from an ocean, as cited by Russell?
2. According to the author, what do scientists believe is the
   cause of "white pox" that is killing coral reefs off Key West,
   Florida?
3. What steps are cities, states, and countries taking to protect
   their coasts, harbors, and oceans, according to Russell?

Excerpted from Dick Russell, "Where the Land Meets the Sea," E/The Environmental
Magazine, March/April 1998. Reprinted with permission from E/The Environmental Magazine;
Subscription Dept.: PO Box 2047, Marion, OH 43306; telephone: 815-734-1242.
Subscriptions are $20 per year.

Rachel Carson, best known for exposing the dangers of DDT in her classic Silent Spring, was also a keen observer of our coasts. "The shore is an ancient world, for as long as there has been an Earth and sea, there has been this place of the meeting of land and water," she wrote in her 1955 book, The Edge of the Sea.

In principle, millions of us feel similarly. Some two-thirds of the Earth's population now live within 50 miles of the sea; nine of the world's 10 largest cities are in coastal zones. U.S. coasts support a third of our national employment—more than 28 million jobs. At the same time, about two-thirds of all the fish caught for our consumption spend their early lives in ecosystems where ocean meets land. In America, commercial and recreational fishing in near-shore waters is a $65 billion industry.

Yet with frightening rapidity, what Carson calls the "intricate fabric of life" is being devastated by human activities. The onslaught is hydra-headed: pollution from industrial and household wastes and agricultural run-off; massive fish die-offs caused by oil spills and the cooling-water intakes of electrical plants; diversions of freshwater from dams; global warming and ozone layer depletion. The tragic fact is, coastal habitat is disintegrating at a rate unprecedented in recorded history.

"Some of the pollution issues are reversible," says Ken Hinman of the National Coalition for Marine Conservation. "There are a lot of examples where we've been able to clean up and restore waterways. The ocean is resilient enough to come back. But the biggest threats are the irreversible changes to the coastline—the filling of wetlands and destruction of barrier islands. How do we reverse the tide of development?" To get a clearer picture of how urgent this situation is, let's take a look back at the events of 1997:

CORAL REEFS

Nearly one-third of the Earth's fish species live along these multi-tiered biological marvels which cover about 160,000 square miles in tropical and subtropical regions. In Asia, the overwhelming majority of species lives on or near reefs, which coastal fishermen rely upon to feed their families. Besides providing shelter and breeding areas, reefs also serve to protect coastlines from storm damage and beach erosion. According to a global study by Australia's Institute of Marine Science in 1992, 10 percent had already degraded "beyond recognition," another 30 percent were in "critical condition," and yet another 30 percent were likely to disappear altogether within the next several decades.

Since that report came out, things have taken an abrupt turn for the worse. Between January and August, 1997, "rapid wasting disease" wiped out huge patches of coral in a 2,000-mile swath ranging from Mexico to the Caribbean. As far away as the Philippines, reefs were being struck by a variety of diseases not seen before. "The problems are occurring at all depths, and the number of species affected is increasing," says Dr. James Porter, a marine ecologist at the University of Georgia, who discovered a mysterious epidemic labeled "white pox." The outbreak was observed February in 1997, in the waters off Key West, Florida, along the only living coral reef in the continental U.S. In some areas, it appears to have eliminated 80 percent of the marine animals. And this is only the latest malady to strike there, following in the wake of "black band" and "white plague II."

And there are no sewage treatment plants in the Florida Keys. According to many scientists, the likely cause of the reef's woes is nutrient pollution from thousands of septic tanks and the untreated outflows from dozens of hotels and resorts. . . .

RED TIDES AND THE "CELL FROM HELL"

Scientists term them "HABs," short for Harmful Algal Blooms, the growth and accumulation of microscopic marine plants that cause "red tides." These blooms are extremely toxic, both to the marine life feeding upon them and to people who eat contaminated shellfish. Since the late 1980s, the number of such single-celled algal species has soared from 22 to 55 around the globe. Outbreaks once found only around certain coastal areas of Europe and the U.S. have spread to Asia and Latin America. Some of this is naturally occurring, as ocean currents deposit seed populations. But algal species are also transferred by the ballast water from ships, and appear to flourish when pollutants add excessive nutrients to the water.

"Red tide off Florida's West Coast comes back nearly every year," says scientific expert Donald Anderson of the Woods Hole Oceanographic Institution. "Large expanses of coastline are closed because of toxins in shellfish. It's eliminated 10 percent of the endangered manatee population, driven countless tourists from beaches, and killed millions of fish."

In 1997, Texas imposed a nearly statewide ban on harvesting oysters, clams and mussels after a red tide killed millions of small fish. Around the same time, panic ensued when a microbe, Pfiesteria piscicida, appeared in several rivers of the Chesapeake Bay system in Maryland and Virginia. Not only did the so-called "cell from hell" kill thousands of menhaden baitfish, it caused severe rashes, flu-like symptoms, and even memory loss in humans who came into contact with it.

Pfiesteria was first detected by North Carolina botanist JoAnn Burkholder in 1988, after mysterious fish kills occurred in local rivers. Burkholder found that the organism thrived in high-growth population areas, correlating to phosphate levels around sewage outfall pipes. Since then, run-off from hog and chicken farms has been identified as another probable cause.

According to the Washington-based Coast Alliance, "Fertilizer and manure, along with more than a billion tons of eroded soil, run off farm fields into coastal waters every year. Almost 30 million pounds of pesticides are annually applied in areas that drain into the nation's coasts.". . .

TURNING THE TIDE?

But amid all of this grim news, there are some hopeful signs. An International Coral Reef Initiative was formally adopted by many nations in 1995, with Australia's Great Barrier Reef Marine Park becoming a model for similar programs. Thailand brought in experts from the University of Rhode Island's Coastal Resources Center to form a management plan for Phuket Island, where new tourist resorts are now required to build on-site sewage treatment plants and reduce sediments washing into the waters. The entire country has since adopted a National Coral Reef Strategy.

Environmentalists are beginning to pressure international lending institutions like the World Bank to cancel their financial support for massive dam-building projects like the under-construction Three Gorges Dam in China, scheduled to be the largest hydroelectric dam in the world by 2009. Such projects are responsible for altering freshwater flows and thus decimating many fish species that depend upon rivers for spawning; half of the original Pacific salmon runs on our West Coast have been lost due to dams, and many others are threatened.

The amended Magnuson-Stevens Fishery Conservation and Management Act passed by Congress in 1996 ordered the National Marine Fisheries Service (NMFS) and its eight regional councils to describe and identify "essential fish habitat" in each of 39 existing fishery management plans. This, according to the National Coalition for Marine Conservation's Hinman, "finally gives fishermen and managers a stronger voice when it comes to protecting habitat."

The Boston Harbor offers one encouraging example of an on-going and successful clean-up. Since dumping of untreated sludge and grease into the harbor was stopped in 1991, water clarity and oxygen levels have noticeably improved. New pumping stations have cut pollution from sewer overflow pipes by 70 percent. Bluefish, striped bass and cod have returned in substantial numbers; harbor seals have been seen swimming near the New England Aquarium there.

The Rhode Island coast has been the victim of two damaging oil spills over the past decade, with the one in 1996 killing more than one million lobsters. Now the federal government is using $270,000 from a settlement with one tanker owner to build artificial stone reefs where lobsters can hide. "What we are trying to do is mimic the natural ecosystems of Narragansett Bay," says project director John Catena.

OUR ONLY OCEAN

We all know by now how blue our earth looks from space—all that water, three-quarters of the surface. Perhaps it's harder to remember that most of our oceans' productivity lies in the coastal regions (90 percent of fisheries harvest takes place within 200 miles of shore, and most of that within the nearest five miles) and that it's those same coastal areas we abuse most with our dumping and runoff and development. In the past century we've disproved Byron's poetic wishfulness about the scope of our damage just as surely as we've disproved the flat-earthers, who thought that the sun sank with a hiss into the infinite ocean at the end of each day. There's not so much ocean out there that it can supply all we demand or suffer every mistreatment we choose to heap upon it. What we ruin no longer stops at the shore. In this International Year of the Ocean [1998], we might ask ourselves not what we can get away with but whether we're willing to commit ourselves to life all the way around our one watery, interflowing globe.

Nancy Lord, *Sierra*, July/August 1998.

The Coast Alliance, a nonprofit organization formed in 1979 to combat development pressure and pollution, is today working with more than 300 state and local groups. One of the Alliance's primary missions is expansion of the Coastal Barrier Resources System, established by Congress in 1982. The system currently includes 1.27 million acres of underdeveloped beaches, dunes, wetlands and barrier islands. While not prohibiting new development altogether, the program eliminated long-standing federal subsidies for private construction in these fragile ecological areas. This has begun to put a significant crimp in beachfront building, although in 1996 a coalition of Florida landowners and developers succeeded in getting Congress to delete 70 acres from the system.

More than 100 efforts in integrated coastal zone management have sprung up in recent years around the globe. Pilot projects are underway in Asia; six East African nations have agreed to make such management government policy. Marine reserves, first established in the 1970s, now include coral reefs, kelp forests, mangroves, salt marshes, sea grass meadows and subtidal/intertidal habitats. Yet, as Michael Berrill points out in his 1997 book *The Plundered Seas*, "Conflicting pressures of the competing commercial interests in the region make coastal zone management slow, difficult and usually ineffective."

THINKING LOCALLY

Clearly, action must start at the local level. One heartening example is the little island of Bonaire off the Venezuelan coast. In addition to adopting some of the first restrictions regarding discharge of ballast water into its harbors (this must happen at least 12 miles out), Bonaire has banned anchoring around its coral reefs and adopted a series of strict rules for divers and tourists.

Faced with collapsing fisheries, residents of a Philippine village called Bindoy formed a group of volunteer Sea Watchers in the late 1980s. By 1993 the group planted more than 100,000 mangrove trees and had sunk more than 1,000 artificial reefs made of bamboo, tires or concrete.

If Bonaire and Bindoy can do it, why is the United Nations Environment Program's plan for "Protection of the Marine Environment from Land-Based Activities" moving so slowly? Solving the crisis in our oceans will obviously take a coordinated worldwide effort, but the bright promise of the 1992 Rio conference is not being fulfilled. The oceans know no borders, and cooperation between nations (now reaching new lows) is essential if the negative trends are to be reversed. Otherwise, the evocative words of Rachel Carson about the magical place where land meets sea will soon be but a distant memory, with the gravest of consequences for all of us.

"In city after city . . . once-fetid
rivers have become focal points for
urban renewal. . . . Cities today view
their waterways as assets, not smelly
nuisances."

WATER POLLUTION IS DECLINING

Gary Turbak

The Clean Water Act of 1972 has dramatically reduced the pollution levels in America's rivers, lakes, and bays, maintains Gary Turbak in the following viewpoint. Cities and industries were forced to stop dumping untreated sewage, pesticides, chemicals, and other pollutants into waterways, he asserts. As a result of the Clean Water Act, Turbak contends that many waterways are the cleanest they have been in years. Turbak is a freelance writer from Missoula, Montana.

As you read, consider the following questions:

1. What were some of the methods that cities and industries used to clean up their waterways, according to the author?
2. What waterways have made a comeback due to the Clean Water Act, according to Turbak?
3. What problems do cities with clean and vibrant waterways now face, in Turbak's opinion?

Reprinted from Gary Turbak, "Streams of Conscience," *The American Legion Magazine*, September 1997, by permission of the author.

On June 22, 1969, the Cuyahoga River burned. Great tongues of flame roared 50 feet into the air. Wooden railroad bridges became torched ruins. Incredulous citizens—and fire fighters, too—stood helplessly on the banks and watched. No one had ever before seen a river burn.

Flowing through the heart of Cleveland, the Cuyahoga had become a moving oil slick upon which growing rafts of debris had long since replaced boaters. Wildlife had abandoned its banks. Sewage and chemicals permeated its currents. And that June a stray spark set the river ablaze.

CESSPOOLS AND SEWERS

The Cuyahoga's flames eventually died down, but the outrage did not. Across the country, citizens looked at their own rivers, lakes and harbors and realized that many of America's once pristine waterways had evolved into open cesspools of neglect. Precious natural resources had become little more than gigantic public toilets.

The Connecticut River, flowing through New England, earned the sobriquet "prettiest sewer in the nation." Its color varied daily, depending on the hue of dye emanating from upstream paper plants, and fetid offal boiled to the surface on stinking bubbles of gas from decomposing sludge. Also in New England, fumes from the polluted Nashua River turned houses black and kept residents awake at night. People joked that birds could walk across the Nashua on the floating sludge, and the city of Fitchburg, Massachusetts, once considered—perhaps only half in jest—capping the river with a parking lot.

In 1965, President Lyndon Johnson called the Potomac a "national disgrace" because of the abundant sewage it carried. In Idaho, the Boise River sometimes ran red with slaughterhouse waste. Lake Erie, screamed the headlines, "is dead." In St. Paul, Minnesota, the Mississippi River ran sewage-filled and mostly fishless between banks lined with auto scrap heaps, industrial plants and train yards. Sea life of all sorts began disappearing from the once fertile Chesapeake Bay.

"Enough!" cried the American people, and in 1972 Congress passed the Clean Water Act, a watershed piece of legislation if there ever was one. Many states followed with their own statutes, and gradually the polluted water paradigm began to change. Reforestation of slopes reduced erosion in many watersheds. Restrictions on industry cleaned up factory effluents. Elimination of pesticides made waters less toxic. And a boom in the construction of municipal waste water treatment plants (from serving 42 percent of the population to serving 74 percent) put sewage in its place. In all, American businesses and taxpayers have since 1972 spent nearly $600 billion cleaning up our waters.

A Dramatic Turn-Around

The result has been a renaissance of water clarity that stretches from Puget Sound to Tampa Bay. The once severely polluted and largely lifeless Boston Harbor now supports bass and bluefish, and porpoises and seals delight cruise ship passengers. For the first time in a decade, swimmers in 1996 returned to the lower Hudson River in New York, a stream that in the 1960s ran raunchy with raw human sewage, industrial chemicals and agricultural runoff.

Aquatic grasses, habitat for a variety of sea life, now thrive in Chesapeake Bay, where striped bass have made a remarkable resur-

---

Chemical Pollution: Ups and Downs

In 1995, the National Oceanic and Atmospheric Administration monitored mussels and oysters for chemical pollutants at 154 sites along the nation's coastline. It found increases in chemical pollutants at 41 sites and decreases at 217 sites.

	Chemical	Number of sites where pollutants increased	Number of sites where pollutants decreased
Pesticides	Chlordane	0	43
	DDT	0	24
	Dieldrin	0	19
Trace Elements	Cadmium	3	20
	Copper	5	17
	Arsenic	5	14
	Selenium	2	12
	Lead	7	8
	Mercury	7	8
	Zinc	6	7
	Nickel	4	5
Other Chemicals	PCB (electrical industry chemical)	0	26
	Butyltin (paint additive)	0	11
	PAH (industrial byproduct)	2	3

National Oceanic and Atmospheric Administration, "Trends in Chemical Concentrations in Mussels and Oysters Collected Along the U.S. Coasts from 1986 to 1993," 1995.

gence. Today, the St. Paul stretch of the Mississippi is cleaner than at any time in the past century and home to 30 species of fish.

Bass—not offal—travel the Potomac, giving our capital a river worthy of pride. The Connecticut River hosts fishing tournaments and rowing competitions, and eagles and ospreys have returned to its banks. In Boise, the greenbelt of riverfront park land has become the city's pride, and summer water recreationists abound.

Massachusetts lists the Nashua, with its teeming fish populations and abundant terrestrial wildlife, as a scenic river. Even Lake Erie is rising from the dead. Fish are returning to some formerly lifeless harbors, and once-common algal blooms are rare. Many people swim and boat, making Lake Erie tourism an $8.5 billion annual business. And the hottest things on the Cuyahoga River these days are popular new shoreside restaurants where pleasure boats vie for docking spots.

In city after city—Hartford, Peoria, Portland (Oregon) and Chattanooga (Tennessee), to name a few, once-fetid rivers have become focal points for urban renewal. Communities these days turn toward—not away from—their rivers as they build parks, theaters and convention centers. Cities today view their waterways as assets, not smelly nuisances, and downtown waterfronts are once again the focal point for urban commerce.

All this, of course, creates a new challenge—how to handle the multitudes of swimmers, boaters, floaters, fishermen, shoppers and others streaming toward America's cleaner waters. But that's the kind of problem any nation would love to have.

PERIODICAL BIBLIOGRAPHY

The following articles have been selected to supplement the diverse views presented in this chapter. Addresses are provided for periodicals not indexed in the *Readers' Guide to Periodical Literature*, the *Alternative Press Index*, the *Social Sciences Index*, or the *Index to Legal Periodicals and Books*.

Nanette Blanchard	"The Quietest War," *E: The Environmental Magazine*, March/April 1998.
Sophie Boukhari	"Marine Blues," *Unesco Courier*, July/August 1998.
William J. Broad	"Survey of 100 U.S. Coastal Sites Shows Pollution Is Declining," *New York Times*, January 21, 1997.
Michael Fumento	"Polluted Science," *Reason*, August/September 1997.
Robert M. Goldberg	"EPA to Asthmatic Kids: Hold Your Breath," *Wall Street Journal*, September 19, 1997.
Thomas D. Hopkins	"Proof? Who Needs Proof? We're the EPA!" *Wall Street Journal*, May 21, 1997.
Nicholas D. Kristof	"Across Asia, a Pollution Disaster Hovers," *New York Times*, November 28, 1997.
John Merline	"How Deadly Is Air Pollution?" *Consumers' Research Magazine*, February 1997.
Steven J. Milloy and Michael Gough	"The EPA's Clean Air-ogance," *Wall Street Journal*, January 7, 1997.
Kiernan Mulvaney	"A Sea of Troubles," *E: The Environmental Magazine*, January/February 1998.
Richard Shedenhelm	"Are We Burying Ourselves in Garbage?" *Freeman*, April 1995. Available from the Foundation for Economic Education, Inc., Irvington-on-Hudson, NY 10533.
T.H. Watkins	"Pollution Can Save Your Life," *New York Times*, June 17, 1997.
Carol Kaesuk Yoon	"A 'Dead Zone' Grows in the Gulf of Mexico," *New York Times*, January 20, 1998.

DO CHEMICAL POLLUTANTS POSE A HEALTH RISK?

Chapter Preface

Farmers in Tifton, Georgia, were distraught when more than 1,000 acres of their peanut crop died after a lime additive had been applied to their fields. They discovered that the fertilizer they had used had had chemical waste from nearby steel mills—with high concentrations of lead and zinc—purposefully mixed in with the limestone. The farmers were even more upset to learn that recycling chemical waste into fertilizer is legal and totally unregulated by the Environmental Protection Agency.

Those who oppose the recycling of industrial chemical waste and sludge into fertilizers contend that the heavy metals are absorbed by the plants, which are then passed up the food chain when the plants are eaten by livestock or humans. Spreading fertilizer enhanced with chemical waste or sludge on crop fields or pastures is "a recipe for catastrophic disruption of the chains of food and life," maintains Abby Rockefeller, the founder and president of ReSource Institute for Low Entropy Systems, a nonprofit organization that works to protect the public health and environment. Most cases of lead poisoning, the recycling opponents point out, are from eating contaminated fruits and grains.

However, supporters of recycling chemical waste into fertilizer argue that the program's benefits outweigh the risks. They assert that companies and the farmers both save money by using recycled chemical waste as fertilizer, that the life expectancy of landfills is extended because the waste that would have once been dumped in them now has another use, and that the fertilizer does help crops grow. According to Charlie Mitchell, a soils specialist from Auburn University in Alabama:

> The farmer is coming out a little ahead. The person spreading it is getting his profit. The company is using its waste instead of dumping it. So we're helping the environment. We're creating jobs. If it's done right, it can really be a win-win situation.

Although scientists do not agree on the risks or dangers of using chemical waste as filler in fertilizers, most scientists do agree that more studies are needed on this controversial issue. The authors in the following chapter debate whether other forms of chemical waste pose a health hazard.

VIEWPOINT 1

"A group of chemicals known as endocrine disrupters and hormone mimickers are undermining the health and genetic viability of hundreds of species, including humans."

CHEMICAL POLLUTANTS ARE A SERIOUS THREAT TO HUMAN HEALTH

Pratap Chatterjee

Toxic chemicals are being found in the environment, in animals, and in humans far from where the chemicals originated, contends Pratap Chatterjee in the following viewpoint. These chemicals are hazardous to human health because studies have linked the pollutants to cancer, birth defects, and other health problems. Chatterjee argues that the only way to ensure the health and survival of humans is to ban all chemicals whose safety has not been proved. Chatterjee is the global environment editor for Inter Press Service based in California.

As you read, consider the following questions:

1. Why are humans particularly vulnerable to the effects of toxic chemicals, in the author's opinion?
2. Why are some chemicals with weak endocrine disrupting abilities particularly dangerous, according to research cited by Chatterjee?
3. In the author's opinion, what will be the result if nothing is done to remove toxic chemicals from the environment?

This article is adapted from Pratap Chatterjee, "Who Is Stealing Our Future?" CAQ (Covert Action Quarterly) #58, Fall 1996, by permission of CAQ, 1500 Massachusetts Ave. NW, Washington, DC 20005, USA. Annual subscriptions in the U.S., $22; Canada, $27; all other areas, $35. The issue of CAQ containing the full text of the article with footnotes is available from CAQ for $8 (add $4 outside the U.S.).

Industry's Toxic Addiction to Estrogen Mimickers and Endocrine Disrupters

Polar bears in the Arctic circle and albatrosses in the middle of the Pacific were the last creatures that scientists expected to be threatened by synthetic chemicals. But the pristine wilderness and the pure ocean vastness are as extinct as the dodo—and just as much casualties of human activity. When the albatross population suffered a 3 percent drop in reproduction rates over the last few years, New Zealand researchers discovered abnormally high levels of synthetic chemicals in the birds' bodies. When polar bear reproduction dropped by more than half, Norwegian researchers documented levels of toxic chemicals in the animals that are 3 billion times higher than in the cold waters near which they live.

The recently published book, Our Stolen Future, brings together mounting scientific evidence that thousands of synthetic chemicals in common use are accumulating all along the food chain and are turning up everywhere from remote virgin forest to supermarket shelf. If the authors are right, a group of chemicals known as endocrine disrupters and hormone mimickers are undermining the health and genetic viability of hundreds of species, including humans. And because the implicated chemicals—including PCBs, chlorine, atrazine, DDT, and various plastics used to manufacture five-gallon water containers and approximately half the canned goods in this country—are so widely used in agriculture and industry, the financial vitality and survival of many corporations is also at stake. Not surprisingly, then, in addition to calls for further investigation and research, the storm of controversy around the new studies implicating these chemicals has also sparked a counterattack funded and promoted by the corporations that would be affected by regulation or a ban.

Mugging the Messengers

The way these chemicals work is to "mimic" or "block" estrogen and progesterone—natural chemicals known as hormones which instruct the body in how it should develop and reproduce. "Hormonally active synthetic chemicals are thugs on the biological information highway that sabotage vital communication. They mug the messengers or impersonate them. They jam signals. They scramble messages," write the authors of Our Stolen Future. For example: "Imagine what would happen if somebody disrupted communications during the construction of a large building so the plumbers did not get the message to install the pipes in half the bathrooms before the carpenters closed the building."

Now, the authors write, imagine that the chemicals that affect communications in the endocrine system are everywhere "in the finest caviar, in penguins in Antarctica, in the bluefin tuna served at a sushi bar in Tokyo, in the monsoon rains that fall on Calcutta, in the milk of a nursing mother in France, in the blubber of a sperm whale cruising the South Pacific."

Throw in a couple more alarming facts. Billions of pounds of these chemicals are pumped annually into the air, land, and water, but the amount required to disrupt reproduction cycles could be as low as one part in a trillion—equivalent to just one drop of liquid in the cars of a six-mile-long cargo train. Humans are particularly vulnerable since the concentration of many of these chemicals increases in animals high in the food chain. The reason is two-fold: First, the chemicals are "persistent," meaning they do not break down, and second, they are stored permanently in body fat so that when a larger animal eats smaller animals, the predator incorporates the pollutants of its prey.

Finally, perhaps the most devastating news of all is that some of the chemicals with weak endocrine disrupting effects on their own become far more dangerous when two or more of them are found together. Research conducted by two scientists from Tulane University in Louisiana—on four pesticides (chlordane, dieldrin, endosulfan, and toxaphene) and several different kinds of PCBs—showed that two or more such chemicals in combination could be as much as 1,600 times as powerful as the individual chemicals alone.

CHEMICAL CATASTROPHE

In the past, scientists looking for the harmful effects of chemical contamination have tended to focus on cancer. While there is evidence linking this class of chemicals to the 32 percent rise in breast cancer rates and the 126 percent increase in prostate cancer in recent years, the situation is more complex and far more alarming. "Humans in their relentless quest for dominance over nature may be inadvertently undermining their own ability to reproduce or to learn and think," warns Our Stolen Future coauthor Theo Colburn. Exposure to estrogen mimicking or endocrine disrupting chemicals such as dioxin may not kill, but may, notes an EPA report, lead to "complex and severe effects including cancer, feminization of males and reduced sperm counts, endometriosis and reproductive impairment in females, birth defects, impaired intellectual development in children, and impaired immune defense against infectious disease."

These chemicals could also be a significant factor in the rapid disappearance of many species around the world, such as the golden toad in Costa Rica, panthers in the Florida Everglades, otters in England, and dolphins off the coast of Turkey. For example, after Tower Chemical spilled large quantities of dicofol, a pesticide closely related to DDT, into Lake Apopka, Florida, in the early 1980s, alligators started appearing with penises so shrunken they could not reproduce.

For fairly obvious reasons, though, the area which has galvanized the scientific community and the media is the link between these chemicals and a well-documented and dramatic drop in human sperm count around the world. Some 61 studies collected by Danish researchers have shown that sperm counts in a number of European countries have fallen by half in the last 30 years, while those in rapidly industrializing countries in East Asia are dropping fast.

MIGRATING POLLUTANTS

Because many persistent organic pollutants (POPs) are semivolatile (i.e., they vaporize at warmer temperatures and condense as the air gets cooler), they can travel long distances on air currents before returning to the earth. POPs travel like grasshoppers, rising into the air, settling back to earth, rising again, moving on air currents. By this means, POPs are "distilled" and they tend to move from warmer climates to colder climates. POPs can also be transported by ocean currents and through the migration of species that carry them in their bodies. Thus the Nordic countries, Canada, Alaska, and other near-Arctic territories are significantly contaminated with hormone-disrupting POPs even though the sources of such chemicals lie thousands of miles to the south.

Peter Montague, *Rachel's Environment and Health Weekly*, June 4, 1998.

DES (diethylstilbestrol) provided one of the first confirmed examples of how these chemicals can affect not only those who are directly exposed, but also future generations. In the late 1950s and '60s this estrogen mimicker was prescribed to millions of women for a variety of problems. Grant Chemicals, one of the manufacturers, claimed that DES produced "bigger and stronger babies," while doctors handed it out to prevent miscarriages, suppress milk production, and as a "morning-after" contraceptive. It was not until the 1970s that researchers discovered that the drug dramatically increases chances of clear-cell cancer and severe damage to the reproductive tract that can result in ectopic pregnancies. (Pregnancies that develop in the fallopian tubes as opposed to the uterus can cause ruptures leading to severe bleeding and sometimes death.) DES is now suspected of having affected male offspring, and of possibly causing brain problems in children of both genders.

INDUSTRY FIGHTS BACK

As they did when faced with evidence of the dangers of DES, tobacco, global warming, nuclear waste, and pesticides, industry leaders have denied that there are any problems, and mounted PR campaigns. Faced with a growing body of evidence on the impact of chemicals on the endocrine system, they have turned to industry-sponsored groups and scientists to disprove the studies available to potential litigants and quoted by environmental groups pushing for regulation. . . .

LIFESTYLE CHANGES

When health authorities and governments make a conscientious effort to set safety standards, they face considerable difficulties. One of the main problems is that the "safe" levels for chemicals in emissions and in everyday products such as pesticides have been traditionally based on their impact on adults, not children, who are at a far greater risk; the assumption is that it is mostly adults who use these products. But there is growing worry that the quantity of the chemical is largely irrelevant; the crucial question is not how much, but when exposure occurs. Thus one part in a million of a certain chemical may be perfectly safe during 99.99 percent of the life-cycle of a normal human being, but exposure to one part in a trillion at a particular time during pregnancy may cause a life-long tragedy.

Given this danger, some activists say the only way to prevent widespread sickness and disease is to question the current course of human "progress." Peter Montague, who has been tracking the effects of synthetic chemicals on human health for 10 years, advocates questioning the use of all such substances. "The studies show that the strange new chemicals that govern our current patterns of lifestyle and consumption are killing us and making us sick," he says. "There is a clear pattern in our history that shows that every time we discover a dangerous chemical, we substitute it with a different one that we know very little about. We can't continue to do this. We have to stop using these chemicals and start living simpler lives."

Some institutions have already suggested that entire classes of chemicals be banned. Studies by the International Joint Commission, a scientific body set up to study water quality in the Great Lakes in Canada and the US, have shown that of the toxic substances found in the lakes, half of those that cause cancer and other health problems contain chlorine. As a result, the Commission recommended phasing out all chlorine-based chemicals. This conclusion was endorsed by the American Public Health Association.

While most scientists and government agencies are taking a "wait and see" approach, some local communities around the country are organizing to get answers for themselves. Last year a grassroots group of women in Marin County, California, a region that has the highest rate of breast cancer in the nation, decided to stop waiting for the medical community and commissioned its own research. The Marin Breast Cancer Watch is currently preparing a survey of the county to try to determine if environmental causes can explain the high cancer rates.

In Seattle, groups including the Women's Health Action Network and the Washington Toxics Coalition meet monthly to talk about issues of reproductive health and synthetic chemicals. Major environmental organizations like the Environmental Defense Fund and Greenpeace have also begun to lobby government and industry on these matters in national capitals.

No Time to Wait

While industry claims we don't know enough to justify action, many activists and researchers warn that if we wait for definitive answers, it may be too late. The cost of doing nothing will be illness and death for individuals, devastation of the environment, and serious genetic damage for many species, including humans. Many of the estimated 100,000 chemicals on the market today have not undergone rigorous testing and about 1,000 new ones are added every year. The burden of proof must shift so that the individual and combined impact of these chemicals is assessed and those that are not proven safe are banned. A phaseout period may be necessary to find natural substitutes and alternatives for substances already in use, but the ultimate goal must be a ban on such substances. In addition, no new chemicals should be introduced until complete testing is completed.

| "The great majority of naturally occurring and synthetic chemicals in the diet appear to be present at levels below which any significant adverse biological effect is likely."

CHEMICAL POLLUTANTS ARE NOT A SERIOUS THREAT TO HUMAN HEALTH

Jonathan Tolman

In the following viewpoint, Jonathan Tolman reports on a study that found that synthetic chemicals do not bear sole responsibility for causing cancer or disrupting the human endocrine system. Many naturally occurring chemicals are present in the human diet and are carcinogenic or act as hormone mimicking chemicals, he maintains. Cancer is more often the result of lifestyle choices rather than chemicals in the diet, he argues. Tolman, an environmental policy analyst at the Competitive Enterprise Institute, is the author of a CEI study, "Nature's Hormone Factory: Endocrine Disrupters in the Natural Environment."

As you read, consider the following questions:

1. What conclusion did Bruce Ames reach concerning the cancer threat posed by food, as cited by Tolman?
2. In what ways are synthetic and naturally occurring chemicals similar, according to the study by the National Academy of Sciences, as cited by the author?
3. According to the NAS, what percentage of cancer cases can be directly attributed to carcinogens in the diet?

Reprinted from Jonathan Tolman, "Rachel Was Wrong," *CEI Update*, March 1996, by permission of the Competitive Enterprise Institute.

Rachel Carson's *Silent Spring* ushered in an era of national concern over the potential effects of synthetic chemicals. Published in 1962, Carson's book suggested that the human use of synthetic chemicals amounted to a "relentless war on life" and that modern society was "losing the right to be called civilized." Americans experienced widespread angst over the potential carcinogenic effects of chemicals over the subsequent decades. As it turns out, Rachel Carson was wrong.

NATURALLY OCCURRING AND SYNTHETIC CHEMICALS

In February 1996, the National Research Council, the research arm of the National Academy of Sciences, released a report on carcinogens in the human diet. Over thirty years after *Silent Spring*'s publication, a wealth of scientific evidence suggests that many of the concerns Carson raised were unfounded.

Plants have evolved numerous chemicals that serve as defensive agents against predators. Some kill predators outright, others act as deterrents in some fashion. Many of these substances can be considered natural pesticides, and are quite common. These chemicals are likely present in our diet in amounts exceeding the residues of synthetic pesticides.

Bruce Ames, of the University of California at Berkeley, began to point out that many naturally occurring chemicals tested positive for carcinogenicity in lab tests on rodents. In fact, the percentage of naturally occurring chemicals identified as "carcinogenic" in rodent bioassays does not differ significantly from that of synthetic chemicals. This led Ames to the conclusion that insofar as there was a cancer risk in the human diet, it was more likely the result of naturally occurring chemicals than synthetics. There are after all, many more naturally occurring chemicals than synthetic. More importantly, Ames posited that the cancer threat posed by food was likely small to nonexistent. The NAS report confirmed the Ames thesis.

The NAS study concluded that based upon existing exposure data, the great majority of naturally occurring and synthetic chemicals in the diet appear to be present at levels below which any significant adverse biological effect is likely. So low, in fact, that they are unlikely to pose any appreciable cancer risk whatsoever.

The study also concluded that natural components of the diet may prove to be of greater concern than synthetic components with respect to cancer risk. As shown in the accompanying chart, synthetic chemicals account for a tiny fraction of the daily dietary intake of substances that have been labeled carcinogens in lab tests. Because of the greater abundance of naturally occurring substances in the diet, the total exposure to naturally occurring carcinogens far exceeds the exposure to synthetic carcinogens.

DAILY INTAKE OF DIETARY CARCINOGENS

Food Category	Daily Intake (g)	Assumed Proportion Carcinogenic	Predicted Intake of Carcinogenic Substances (g)
Traditional Foods	1,000	0.001	1
Spices and Flavors	1	0.01	0.01
Indirects	0.02	0.1	0.002
Charred Protein	1	0.0001	0.0001
Animal Drugs	0.001	0.1	0.0001
Pesticides	0.0002	0.5	0.0001
Mycotoxins	0.00001	0.1	0.000001

National Academy of Sciences, "Carcinogens and Anti-Carcinogens in the Human Diet, 1996."

Another finding of the NAS study was that the basic mechanisms involved in the entire process of cancer—from exposure of the organism to expression of tumors—are similar, if not identical for synthetic and naturally occurring carcinogens. The NAS study concluded that there was no notable difference between synthetic and naturally occurring carcinogens. The argument made by many environmentalists that the human body has evolutionary defenses against natural carcinogens, but not against synthetic ones, does not hold water.

DENIAL

Despite the strong scientific consensus that underlies the NAS report, many environmentalists (who have made their careers opposing the trace levels of pesticides on foods) still can't seem to believe it. Al Meyerhoff, senior attorney for the Natural Resources Defense Council, told the New York Times that despite the report, synthetic chemicals "can still cause thousands of cancers in consumers, and they should be avoided whenever possible." Psychologists usually refer to this as denial.

The NAS study reviewed the issue of how many cancers might be caused by carcinogens in the diet. In citing one study the authors acknowledged that "except for alcohol, the known dietary carcinogens could not account for more than a few hundred cancer cases." This amounts to less than 0.01 percent—one ten-thousandth—of all cancers. What actually does cause cancer, if not pesticide residues and other chemicals, is not earth-shattering news. Smoking, excessive alcohol consumption, and overeating were listed as the biggest contributors to an increased cancer risk.

But just when you thought it was safe to go back into the supermarket, environmentalists have created another environmental affliction. Using cancer as a scare tactic appears to be waning, so environmentalists have latched onto a new cause celebré—endocrine disrupters.

ENDOCRINE DISRUPTERS

The new theory is that certain synthetic chemicals, such as DDT, PCBs, various organochlorines and other synthetic chemicals, mimic estrogen or other hormones and can interfere in human development and reproduction. Among the charges made is that these chemicals are responsible for a global decline in human sperm counts—even though there is no conclusive evidence that sperm counts are declining.

This collective hypochondria is laid out in the 1996 book by World Wildlife Fund Senior Scientist Theo Colburn, environmental writer Diane Dumanowski, and W. Alton Jones Foundation Director John Peterson Myers. Their collective angst is ominously entitled, Our Stolen Future.

Unfortunately for the proponents of synthetic endocrine disruption, mother nature has already beaten them to it. Just as they produce carcinogens, a large number of plants routinely produce significant quantities of hormone mimicking chemicals. The human diet is simply filled with foods that contain naturally occurring endocrine disrupters from genestein in soybeans to cafesterol in coffee.

Given the wealth of human exposure to these substances, if there were widespread human effects, the world would have seen them by now. The lack of compelling data suggests that the kids are all right, and there is no cause for alarm. The issue of endocrine disruption clearly warrants further study, but not the sort of hysteria routinely ginned up by environmental activists.

Let's just hope that it doesn't take science 30 years to do for concerns about endocrine disrupters what it did for concerns about cancer.

| "Despite the persistent efforts of industry to detoxify dioxin, the weight of evidence from scientific literature today confirms its pervasive toxic effects."

EXPOSURE TO DIOXIN LEADS TO HEALTH PROBLEMS

Stephen Lester

Dioxin is a chemical that is a toxic by-product of chlorine and other compounds. In the following viewpoint, Stephen Lester asserts that studies by chemical companies to prove that exposure to dioxin is not a health hazard are outright distortions of the truth. Dioxin has been conclusively linked to cancer and disfiguring skin diseases, Lester maintains, and is a pervasive toxic chemical. The theory advanced by chemical companies that dioxin is a naturally occurring toxic chemical is a preposterous attempt to blind the public to its dangers. Lester is the science director for the Center for Health, Environment, and Justice in Falls Church, Virginia.

As you read, consider the following questions:

1. What were the major scientific flaws found in the Monsanto and BASF studies on dioxin and cancer, according to Lester?
2. What is the "combustion theory," as reported by the author?
3. In Lester's opinion, why is the combustion theory not supported by scientific evidence?

Reprinted from Stephen Lester, "Industry's 'True Lies,'" *Everyone's Backyard*, Summer 1995, by permission of the Citizens Clearinghouse for Hazardous Waste.

The full story of dioxin is a complex one, and includes cover-ups, lies, and deceit; data manipulation by corporations and government; and fraudulent claims and faked studies. For the public, it is a story of pain, suffering, anger, betrayal, and rage; of birth defects, cancer, and many uncertainties about health problems.

Although many companies have contributed to the dioxin story, three chemical companies have played particularly significant roles: Monsanto, BASF, and Dow Chemical. All three manufactured commercial products that were contaminated with dioxin. All three conducted health studies to evaluate dioxin toxicity, which were then used for many years to support claims that there were no long-term effects, including cancer, from dioxin exposure.

THE "CLASSIC" DIOXIN STUDIES

In 1949, an explosion at the Monsanto chemical plant in Nitro, West Virginia, exposed many workers to the dioxin-contaminated herbicide 2,4,5-T. Thirty years later, Monsanto scientists and an independent researcher, Dr. Raymond Suskind, compared death rates among workers they said had been exposed to the death rates of workers who were not exposed. When no differences between the two groups were found, Monsanto claimed that dioxin did not cause cancer and that there were no long-term effects from dioxin exposure. Monsanto released additional studies from 1980 to 1984 supporting this general conclusion that there was no evidence of adverse health effects, other than chloracne, in workers exposed in the 1949 accident.

Similarly, a chemical accident in 1953 at a BASF trichlorophenol plant in Germany released dioxin-contaminated chemicals, exposing workers and the nearby communities of Mannheim and Ludwigshafen. Again, scientists working for the company looked at cancer rates nearly thirty years later and reported no differences between workers who were exposed and workers who were not exposed during the accident. Both the BASF and the first Monsanto study were released in 1980, shortly after researchers at Dow Chemical Company found that very low levels of dioxin caused cancer in rats. BASF, Monsanto and others argued that their studies conclusively showed that dioxin did not cause cancer in humans. Industry used the BASF and Monsanto results to challenge the Environmental Protection Agency (EPA) efforts to regulate dioxin as a probable human carcinogen, arguing that humans respond differently to dioxin than do laboratory animals. People must be less sensitive, they argued. Otherwise, some evidence of cancer would have been found in the two "classic" studies. But when both the Monsanto and BASF studies were re-examined, the methodology used in both was found to have serious scientific flaws.

MONSANTO

Evidence of inaccuracies in both the Monsanto and BASF studies was first revealed during the *Kemner vs. Monsanto* trial, in which a group of citizens in Sturgeon, Missouri, sued Monsanto for alleged injuries suffered during a chemical spill caused by a train derailment in 1979. While reviewing documents obtained from Monsanto during discovery, lawyers for the victims noticed that in one of the Monsanto studies, certain people were classified as dioxin exposed, while in a later study, the same people were classified as not exposed.

These documents revealed that Monsanto scientists omitted five deaths from the dioxin-exposed group and took four workers who had been exposed and put them in the unexposed group. This resulted in a decrease in the observed death rate for the dioxin-exposed group, and an increase in the observed death rate for the non-exposed group. Based on this misclassification of data, the researchers concluded that there was no relation between dioxin exposure and cancer in humans.

In truth, the death rate in the dioxin exposed group of Nitro workers was 65% higher than expected, with death rates from certain diseases (such as lung, genitourinary, bladder, and lymphatic cancers, and heart disease) showing large increases.

Another Suskind study did not look at an original group of workers known to be dioxin exposed, but instead looked at hundreds of Monsanto workers at the Nitro facility. Some of the same classification sleight-of-hand was performed in this study. Again, documents uncovered in *Kemner vs. Monsanto* showed that in fact there were 28 cancer cases in the exposed-worker group and only two in the unexposed group. Suskind, however, reported finding only 14 cancers in the exposed-worker group, compared to six in the unexposed group.

Suskind also examined a group of 37 exposed Monsanto workers during the four year period following the 1949 accident. Medical documents obtained by Greenpeace from the Sloan-Kettering Institute in Cincinnati, Ohio, where Suskind worked, showed that workers suffered "aches, pain, fatigue, nervousness, loss of libido, irritability and other symptoms, active skin lesions, definite patterns of psychological disorders." All but one of the 37 workers had developed chloracne, a severe skin condition. But in a report to Monsanto at the time, Suskind concluded, without further explanation, that "his findings were limited to the skin"; in other words, all other health effects of dioxin exposure, besides chloracne, were not reported. Out of these studies grew the industry claim that chloracne is the only long-term effect of dioxin exposure.

BASF

The study of BASF workers exposed to dioxin in 1953 was also found to have serious scientific flaws. BASF workers weren't convinced by company scientists' claim that there was no evidence of any health problems, other than chloracne, linked to dioxin exposure. They hired their own independent scientists to review the data. This review found that some workers who had developed chloracne, known to occur only in people exposed to high levels of dioxin, were included in the low or unexposed group in the study. In addition, the exposed group had been "diluted" with 20 supervisory employees who appeared to be unexposed. When these 20 people were removed from the ex-

Reprinted by permission of John Jonik.

posed group, significant increases in cancer were found among the exposed workers.

In February 1990, Dr. Cate Jenkins, project manager for the EPA Waste Characterization and Assessment Division of the Office of Solid Waste, alerted EPA's Science Advisory Board about the revelations of fraud in the BASF and Monsanto studies. The Board, which is an independent group of scientists from outside the agency, had recently completed a review of the cancer data on dioxin, which included the BASF and Monsanto studies, and concluded that there was "conflicting" evidence about whether dioxin caused cancer in humans. The Board recommended that EPA continue to rely on data from animal studies.

This animal study data however, had been under attack since mid-1987 when, under pressure from industry, EPA stated that they may have "overestimated" the risks of dioxin. The agency was then preparing to weaken their risk estimate for dioxin based largely on the exposure effects reported in the Monsanto and BASF studies.

Jenkins asked EPA to re-evaluate the proposed regulatory changes and to conduct a scientific audit of Monsanto's dioxin studies. Instead, in August 1990, the EPA Office of Criminal Investigations (OCI) recommended a "full field criminal investigation be initiated by OCI." After two years OCI abandoned the investigation because some of the alleged criminal activities were "beyond the statute of limitation." In fact, EPA actually spent two years investigating Cate Jenkins.

Subsequent studies on the exposed workers at both the Monsanto plant and the BASF plant have been published in scientific journals. In 1991, the National Institute of Occupational Safety and Health (NIOSH) re-examined the causes of death in workers at the Nitro plant and found increases in all cancers. Similarly, in 1989, data on the BASF workers was re-examined and an increase in all cancers was found for workers with chloracne and with 20 or more years since exposure. The re-examination of these once "classic" studies provides strong evidence that the workers exposed to dioxin-contaminated chemicals in these two accidents did indeed suffer higher rates of cancer.

THE DOW CHEMICAL COMPANY

Dow Chemical Company produced the herbicides 2,4,5-T and Agent Orange, the defoliant that was sprayed on the jungles of Vietnam. Both herbicides are contaminated with dioxin during the manufacturing process.

In 1965, Dow conducted a series of experiments to evaluate

the toxicity of dioxin on inmates at Holmesburg prison in Pennsylvania. Under the direction of Dow researchers, pure dioxin was applied to the skin of prisoners. According to Dow, these men developed chloracne but no other health problems. But no health records are available to confirm these findings, and no follow-up was done on the prisoners, even after several went to the EPA after they were released seeking help because they were sick. EPA did not help them.

In 1976, Dow began studies to evaluate whether animals exposed to dioxin would develop cancer. Dow chose very low exposure levels, perhaps anticipating that the studies would show no toxic effects at low levels. Much to their surprise, they found cancer at very low levels, the lowest being 210 parts per trillion.

Around the same time, evidence was found of increased miscarriages in areas of the Pacific Northwest that were sprayed with the herbicide 2,4,5-T. Based on these findings, the EPA proposed a ban on the herbicide. Dow brought their scientists to Washington and created enough pressure that by 1979 EPA had decided to only "suspend" most uses of 2,4,5-T. This enabled Dow to continue to produce this poison until 1983, when all uses of the herbicide were finally banned.

In mid-1978, the Michigan Department of Natural Resources found dioxin in fish in the Tittabawassee and Saginaw rivers. Dow discharged wastewater into these rivers from its plant in Midland.

THE COMBUSTION THEORY

Dow responded in a most unusual way. In November 1978, after an intense four and one half month effort that cost the company $1.8 million, Dow released a report called the "Trace Chemistries of Fire," which introduced the idea that dioxin was present everywhere and that its source was combustion and any and all forms of burning. Dow released the report at a press conference rather than in the scientific literature, which is the standard procedure with scientific studies. The report concluded that dioxin in the Tittabawassee and Saginaw rivers came not from Dow, but from "normal combustion processes that occur everywhere." A Dow scientist stated at the time that, "We now think dioxins have been with us since the advent of fire."

Subsequent studies have proven the "combustion theory" claims to be more public-relations myth than scientific fact. Measurements of dioxin in lake sediments show that dioxin levels dramatically increased after 1940, when chemical companies such as Dow began to make products contaminated with dioxin.

Other studies reveal that prehistoric humans, who burned

wood for fuel, did not have significant quantities of dioxin in their bodies. Tissues from 2,000 year-old Chilean Indian mummies did not have dioxin. EPA states in its reassessment that dioxin can be formed through natural combustion sources, but this contribution to levels in the environment "probably is insignificant."

Despite the persistent efforts of industry to detoxify dioxin, the weight of evidence from scientific literature today confirms its pervasive toxic effects. Faced with the toxic truth about the dioxin they create, industry has two choices; either stop producing dioxin, or continue to deliberately poison the public policy debate with lies and conflicting information. History tells us they will continue the lies until we make them own up to the truth.

| "Life-threatening health effects in humans have not been linked definitively to dioxin, despite our fears to the contrary."

EXPOSURE TO DIOXIN DOES NOT LEAD TO HEALTH PROBLEMS

American Council on Science and Health

Dioxin is a chemical by-product that is formed when materials composed of chlorine-based compounds are burned or otherwise deteriorate. In the following viewpoint, the American Council on Science and Health (ACSH) argues that dioxin's health risks are exaggerated. Thousands of people have been exposed to high levels of dioxin with few ill effects, the council maintains. Furthermore, studies comparing groups of workers who were exposed to dioxin to those who were not exposed show little difference in cancer and birth defect rates. Chlorine is an extremely useful compound that must not be banned, the council concludes. The American Council on Science and Health is a consortium of physicians, scientists, and policy analysts who study issues related to health, the environment, and lifestyles.

As you read, consider the following questions:

1. According to the ACSH, what industrial processes produce dioxin?
2. What is the only adverse health effect that can be conclusively linked to exposure to dioxin, as cited by the author?
3. In what ways is chlorine necessary to maintain high standards for food, water, and housing, according to the ACSH?

Reprinted, with permission, from *Chlorine and Health*, a publication of the American Council on Science and Health, 1995 Broadway, 2nd Fl., New York, NY 10023-5860.

Numerous reports in the media have ascribed possible detrimental health effects to chlorine, dioxin and other chlorinated chemicals, often subjecting the public to exaggerated and misleading information. Most of these stories have no basis in scientific fact.

THE CHARGES

Greenpeace, a worldwide environmental activist group, has led the attack, pushing for a total ban on chlorine and chlorinated chemicals. With its document *The Product is the Poison—The Case for a Chlorine Phase-Out*, Greenpeace took the first step in a campaign to shut down the chlorine industry. In that document Greenpeace recommended phasing out "the use, export, and import of all organochlorines, elemental chlorine, and chlorinated oxidizing agents," charging, without scientific justification:

• that the small amounts of dioxins produced during the chlorine bleaching of paper pulp pose an unacceptable risk to the population;

• that polyvinyl chloride (PVC), a chlorine-containing polymeric chemical widely used in consumer products, is "uniquely damaging during production, use and disposal," and that it is a potent producer of dioxins during incineration; and

• that all chlorinated chemicals may be carcinogenic and may have negative reproductive effects.

Shortly after the release of the Greenpeace report and the subsequent media barrage, the Environmental Protection Agency announced, under the authority of the Clean Water Act, the initiation of a 2½-year study with the goal of "develop[ing] a national strategy for substituting, reducing or prohibiting the use of chlorine and chlorinated compounds." Nothing in the EPA's statement indicated that the health effects of chlorinated compounds would be evaluated.

The EPA announcement came on the heels of a major study by the International Joint Commission (IJC), a U.S.-Canadian group that oversees implementation of the Great Lakes Water Quality Agreement. In its "Seventh Biennial Report Under the Great Lakes Water Quality Agreement of 1978" published in early 1994, the IJC recommended severe reductions in chlorine use. The recommendation, which would affect the future use of chlorine and chlorinated chemicals, totally disregarded the scientific data about these chemicals and was not supported by the group's own science advisers, who concluded that there was a need for "a thorough and complete analysis of chlorine chemistry before any schedule for sunsetting chlorine is implemented."

SCIENTISTS RESPOND

In response to these events, several groups of scientists have emphasized that current regulations dealing with chlorinated chemicals are sufficient and that there is no need to ban the entire group. For example . . . an extensive study published in *Regulatory Toxicology and Pharmacology* evaluated the use of chlorine in several industries, including the PVC, pulp and paper, drinking water, incineration and pesticide industries. The 1,100-page study, published in September 1994, concluded that "although much remains to be learned about chlorinated organic chemicals, enough is known to ensure that now and in the future, they can be used and discharged with assurance that adverse effects will be absent.". . .

DIOXIN

The name "dioxin" refers to a family of about 75 chemicals, among them 2,3,7,8-tetrachlorodibenzo-p-dioxin (TCDD), generally considered the most toxic of the group. Dioxins were first synthesized in the late 1950s at Ohio State University. Their toxic effects in animals were noted soon after.

Dioxins have no known use. They are by-products of some industrial processes that use chlorine, such as the bleaching of paper pulp. Dioxins are also produced during any combustion process—waste incineration, running motor-vehicle engines, steel-making and smelting, residential wood burning and even forest fires.

Despite assorted claims over the past 20 years, the dioxin known as TCDD is not the "doomsday chemical of the 20th century," nor is it the "deadliest substance ever created by chemists." The people who have been exposed to high levels of TCDD number in the thousands. Physicians and epidemiologists have been observing the health of those individuals—industrial workers, civilians, Vietnam veterans—who were exposed to TCDD at high levels during the past 40 years. Even after all these studies, described in more detail below, not one has been able to attribute unequivocally any human cancers or deaths to TCDD exposure. The only documented adverse health effect is the skin disease chloracne. Although it is often persistent and disfiguring, chloracne is not life-threatening and is often reversible when exposure ceases.

While dioxin tissue levels among Vietnam veterans in general are not significantly different (11.7 parts per trillion [ppt]) from the levels of non-Vietnam veterans (soldiers who had never been to Vietnam; 10.9 ppt) or a civilian control group (12.4 ppt), certain groups of Vietnam veterans were exposed to higher levels of dioxin, a contaminant of the defoliant Agent Orange. Studies following these individuals show no association between dioxin tissue levels and cancer or other health effects. A two-part, 20-year mortality and health-effects evaluation of 995 Air Force Ranch Hands, the personnel who handled and sprayed Agent Orange, revealed that some had high tissue concentrations of dioxin (>300 ppt) 15 years after exposure. Among this group, there was no chloracne observed, no increase in nine immune-system tests and no increase in either melanoma or systemic cancer (cancers of the lung, colon, testicle, bladder, kidney, prostate; Hodgkin's disease; soft tissue sarcoma or non–Hodgkin's lymphoma). The authors of this 1990 study concluded that "there is insufficient scientific evidence to implicate a causal relationship between herbicide exposure and adverse health in the Ranch Hand Group."

Dioxin Study Results

Studies of more than 800 dioxin-exposed workers in nine industrial-plant accidents in the United States, England, Germany, France, Czechoslovakia and the Netherlands fail to indicate serious long-term health effects in these men, some of whom have dioxin concentrations exceeding 1,000 ppt 30 years after their initial exposure. Some 465 cases of chloracne were observed in these workers.

A study of 2,200 Dow Chemical workers who were potentially exposed to dioxin revealed that they had a slightly lower mortality than a control group and that they have had no total cancer increase. A study of 370 wives of dioxin-exposed men showed no excess miscarriages and no excess fetal deaths or birth defects in their children.

The Institute of Occupational Health at the University of Milan has published detailed evaluations of the human health effects of the July 1976 dioxin accident involving 37,000 people in Seveso, Italy. Some of the exposed children in "Zone A," the area of heaviest exposure, had dioxin tissue levels as high as 56,000 ppt immediately following the accident; but the only adverse health effect has been chloracne. Of the 193 cases of chloracne, 170 were in children under the age of 15; and the skin lesions in all but one of these cases had disappeared by 1985. Although it is essential to continue to monitor the health of the people in Seveso, the Institute report concluded that there were "no increased birth defects due to dioxin exposure," since the children born during the period from 1977 to 1982 failed to demonstrate an increased risk of birth defects.

The Seveso cancer mortality findings after 10 years do not allow firm conclusions. On the one hand, mortality from cancer of the liver, one of the organs targeted by dioxin, was no different from that of unexposed people; and breast cancer mortality tended to be below expectations. On the other hand, "increases in biliary cancer, brain cancer, and lymphatic and hemopoietic cancer did not appear to result from chance. However, no definite patterns related to exposure classification were apparent."

In the words of a leading dioxin analyst, Dr. Michael Gough:

> No human illness, other than the skin disease chloracne, which has occurred only in highly exposed people, has been convincingly associated with dioxin. In short, epidemiologic studies in which dioxin exposures are known to have been high, either because of the appearance of chloracne or from measurements of dioxin in exposed people, have failed to reveal any consistent excess of cancer. In those studies that have reported associations between exposure and disease, no chloracne was reported, and there are no measurements of higher-than-background levels of dioxin in the people who are classified as exposed.

Life-threatening health effects in humans have not been linked definitively to dioxin, despite our fears to the contrary. Over 40,000 scientific papers have provided enormous information about this greatly misunderstood chemical, and the scientific and medical communities will continue to monitor the health of those people who have been exposed to large amounts of dioxins.

ESTROGENIC EFFECTS ON FERTILITY AND BREAST CANCER

Besides cancer, there are other health "endpoints" about which there is concern relating to dioxin and other organochlorines. It has been alleged that dioxin and other chlorinated chemicals mimic estrogen, adversely affecting the immune system and possibly inducing birth defects. Although dioxin is a teratogen (birth-deforming agent) in laboratory animals, none of the many studies undertaken show that dioxin causes birth defects in humans. The most heavily exposed group of women in Seveso showed no increased incidence of birth abnormalities in their newborn children. Moreover, the examination of medically aborted fetuses during the period following the Seveso accident failed to indicate birth defects. Tragically, many women in Seveso needlessly elected to have abortions out of fear that their children would be born with defects.

Various reports linking decreased human sperm counts worldwide to chlorinated chemical exposure are based on questionable data and grand exaggeration. Much of the furor over sperm counts came from a 1992 report in the *British Medical Journal* citing a 50 percent decline in sperm counts from 1938 to 1990 among men from industrialized countries. After promoting, in a 1993 *Lancet* article, the hypothesis that this decline was associated with estrogenic compounds, the authors later admitted that the apparent decrease in sperm counts was due to computational error and was not supported by a reanalysis of the data. A 10-year study of the semen quality of Wisconsin men showed no change over time in sperm concentration or motility. (It must be noted, however, that virtually all studies of sperm health suffer from methodological problems, including how subjects are selected and the number of samples taken.)

NOT TOXIC IN LOW DOSES

Several studies examining humans exposed to dioxin indicate that the chemical is not toxic in low doses. By studying workers from certain chemical plants, industrial accidents, and the "Ranch Hands" (the American soldiers in Vietnam who cleared forests with dioxin-containing Agent Orange), epidemiologists find that rates of cancer and other diseases in dioxin-exposed individuals were comparable to those of the general population.

Julie DeFalco, *Washington Times*, February 22, 1996.

While sperm concentration and motility are not the only determinants of male fertility, the 1965 Princeton National Fertility Study and the large, broad-based surveys conducted by the National Center for Health Statistics in 1976, 1982 and 1988 indicate that rates of infertility have remained constant over the past three decades at 8 to 11 percent, with male infertility accounting for approximately one third of the cases.

Furthermore, reproductive problems have not been detected in the groups of people who were most heavily exposed to dioxin: Vietnam Air Force Ranch Hands, occupationally exposed workers and the populations of Times Beach and Seveso.

BREAST CANCER

Investigations into the causes of breast cancer are another area of controversy in which chlorinated estrogenic chemicals have been controversially linked to disease. A 1993 *Journal of the National Cancer Institute* report found that DDE, a breakdown product of the now-banned pesticide DDT, was present in higher concentrations among a small group of Long Island women with breast cancer when they were compared with a control group. This observation led the researchers to suggest that there is an association between DDT exposure and breast cancer. But a more recent larger study published in the same journal found no association between breast cancer and higher levels of DDE or PCBs [polychlorinated biphenyls were used as coolants and lubricators, but are now banned in the United States]. Without agreement in even a small number of studies, it is highly premature to assume that chlorinated environmental contaminants are influencing breast cancer rates.

It is important to recognize that there is a downward trend in the emissions of chlorinated compounds. Chlorinated pesticides that formerly were associated with adverse effects on wildlife are no longer being used. The concentrations of these chemicals are declining in the environment, and previously affected species are recovering. Industrial and incineration-emission technology has improved dramatically in recent years, leading to decreased emissions of chlorinated compounds and by-products. The EPA has reported that dioxin emissions, which peaked in 1970, have been decreasing since 1980, demonstrating that the regulatory system is working well to protect the environment.

THE BENEFITS OF CHLORINE

Chlorine is an inextricable part of our lives and is necessary for the maintenance of the present high standards in our food, water and housing. Chlorine contributes in the fields of medicine, transportation and communications. Chlorine is a building block for nearly all chemical processes; it plays a vital role in the health of the population and in maintaining a clean and safe environment. From chlorine-containing pharmaceuticals to fire-resistant and recyclable PVC construction materials, and from water purification to the raising of safe, insect-free food crops, chlorine makes a crucial contribution to the health and well-being of our society.

If chlorine and its chemical derivatives are banned, the expense to the American people in finding replacements will cost billions of dollars and result in the loss of hundreds of thousands of jobs. The loss of useful products—plastics, pharmaceuticals, even safe drinking water—will be a needless tragedy. We will have taken a giant step backward in our standard of living.

The benefits—and the adverse consequences—of chlorinated compounds must be carefully evaluated before any attempt is made to ban specific compounds or entire classes of compounds.

New efforts in Congress to evaluate chemicals using risk assessment and to evaluate regulations based on cost/benefit analysis should help in this process. Policymakers and regulators must have an informed understanding of the toxicity of chlorine and of the thousands of chlorinated compounds in use today. They must also have a more informed understanding of both the benefits and the difficulties associated with substitutions and outright bans.

The scientific community in industry, academia and government must continue to ensure that the future use of chlorine, PVC and other chlorinated chemicals is based on sound science, thoughtful risk assessment and cost/benefit analysis, along with full consideration of health and environmental factors.

| "'Organophosphate' insecticides
(OPs) have the potential to cause
long-term damage to the [child's]
brain and nervous system."

ORGANOPHOSPHATE INSECTICIDES
MUST BE BANNED

Richard Wiles, Kert Davies, and Christopher Campbell

Organophosphate insecticides (OPs) kill insects by interfering with their neurological system. In the following viewpoint, Richard Wiles, Kert Davies, and Christopher Campbell argue that the chemicals in some of the most commonly used organophosphate insecticides are also toxic to humans. The authors maintain that the residue left on food from OPs poses a health hazard for humans, especially for children, who, because of their small size, are highly susceptible to adverse effects of these chemicals. The Environmental Protection Agency must protect children's health by banning the use of the most dangerous insecticides, the authors assert. Wiles, Davies, and Campbell are the coauthors of "Overexposed: Organophosphate Insecticides in Children's Food," published by the Environmental Working Group, an environmental research organization in Washington, D.C.

As you read, consider the following questions:

1. According to the authors, how many children age five and under eat an unsafe dose of organophosphate insecticides every day?
2. What one fruit do the authors say is responsible for just over half the unsafe levels of OPs eaten by children each day?
3. In the authors' opinion, why is the situation with children and organophosphate insecticides worse than the earlier circumstances with lead?

Reprinted, with permission, from Richard Wiles, Kert Davies, and Christopher Campbell, "Overexposed: Organophosphate Insecticides in Children's Food," a January 1998 report of the Environmental Working Group.

Every day, nine out of ten American children between the ages of 6 months and 5 years are exposed to combinations of 13 different neurotoxic insecticides in the foods they eat. While the amounts consumed rarely cause acute illness, these "organophosphate" insecticides (OPs) have the potential to cause long-term damage to the brain and the nervous system, which are rapidly growing and extremely vulnerable to injury during fetal development, infancy and early childhood.

CHILDREN ARE AT RISK

Based on the most recent government data available on children's eating patterns, pesticides in food, and the toxicity of organophosphate insecticides, we estimate that:

• Every day, more than one million children age 5 and under (1 out of 20) eat an unsafe dose of organophosphate insecticides. One hundred thousand of these children exceed the Environmental Protection Agency (EPA) safe dose, the so-called reference dose, by a factor of 10 or more.

• For infants six to twelve months of age, commercial baby food is the dominant source of unsafe levels of OP insecticides. OPs in baby food apple juice, pears, applesauce, and peaches expose about 77,000 infants each day, to unsafe levels of OP insecticides.

This estimate very likely understates the number of children at risk because our analysis does not include residential and other exposures to these compounds, which can be substantial, and because EPA's estimates of a safe daily dose (the so-called reference dose or RfD) are based on studies on adult animals or adult humans, and almost never include additional protections to shelter the young from the toxic effects of OPs.

Our analysis also identified foods that expose young children to the most toxic doses of these pesticides. We found that:

• One out of every four times a child age five or under eats a peach, he or she is exposed to an unsafe level of OP insecticides. Thirteen percent of the apples, 7.5 percent of the pears and 5 percent of the grapes in the U.S. food supply expose the average young child eating these fruits to unsafe levels of OP insecticides (Table 1).

• A small but worrisome percentage of these fruits—1.5 to 2 percent of the apples, grapes, and pears, and 15 percent of the peaches—are so contaminated with OPs that the average 25-pound, one-year-old eating just two grapes, or three bites of an apple, pear, or peach (10 grams of each fruit) will exceed the EPA (adult) safe daily dose of OPs.

- The foods that expose the most children age six months through five years to unsafe levels of OPs (because they are more heavily consumed) are apples, peaches, applesauce, popcorn, grapes, corn chips, and apple juice. Just over half of the children that eat an unsafe level of OPs each day, 575,000 children, receive this unsafe dose from apple products alone.
- Many of these exposures exceed safe levels by wide margins. OPs on apples, peaches, grapes, pear baby food and pears cause 85,000 children each day to exceed the federal safety standard by a factor of ten or more.

A COMPREHENSIVE ANALYSIS OF TOXIC DOSES

This Environmental Working Group study utilizes detailed government data on food consumption patterns and pesticide residues to conduct the first comprehensive analysis of the toxic dose that infants and children receive when the entire organophosphate family of insect killers is assessed in combinations, and at levels, that actually occur in the food supply.

The study was prompted by the 1996 Food Quality Protection Act, which requires the government, for the first time, to consider the total risk posed to humans when they are exposed to any and all pesticides that have a common mode of toxic action and a similar type of effect. Prior to 1996 law, the government determined a separate, "safe" level of exposure for each of the dozens of registered pesticides found in food, but did not regulate as a group chemicals that produce the similar health problems. The new law further required specific protections for infants and children, who are more vulnerable to pesticides and other toxins.

Recently, the Environmental Protection Agency (EPA) concluded that the organophosphates have a common toxic mechanism, and that exposure to combinations of the chemicals should be considered in setting a "safe" dose.

FQPA MANDATES EXTRA PROTECTION FOR KIDS

The Food Quality Protection Act (FQPA) requires EPA to act to protect infant and child health, even in the absence of total scientific certainty regarding the toxicity or exposure of pesticides to the fetus, infant or young child. This is a dramatic reversal of previous statutory requirements where EPA had no mandate, and arguably could not act to protect the public health, even child health, in the absence of complete data on the risk from a pesticide. Now the law is clear. In the absence of complete and reliable data on pre- and postnatal toxicity and exposure to a pesticide, the

EPA must err on the side of child safety and apply an additional ten-fold margin of safety to food tolerances for the pesticide.

Contrary to the clear requirements of the law, the EPA has devised and implemented an official policy in response to FQPA that disregards the requirement for a ten-fold safety factor. This policy plainly undermines protection of the nation's children from pesticides. If the new ten-fold safety factor were applied to all organophosphate insecticides found in food, we estimate that nearly 3.6 million children age 5 and under would be exposed to levels of these pesticides in food that would exceed the new standard.

HIGH RISK PESTICIDES

Analysis of more than 80,000 samples of food inspected by the federal government for pesticide residues from 1991 through 1996, revealed that 13 organophosphate insecticides were found in or on food by the Food and Drug Administration and the U.S. Department of Agriculture.

The highest risk OP compounds are methyl parathion, dimethoate, chlorpyrifos, pirimiphos methyl, and azinphos methyl which account for more than 90 percent of the risk from OP insecticides in the infant and child diet.

Achieving a safe food supply for children, however, is not as simple as banning the five highest risk OPs. Home and other non-food uses must be considered, as well as the fact that other OPs will likely substitute for those that are banned in the first wave of standard setting. And most importantly, infant and child safety must be measured in terms of safety standards designed to protect these children, not in terms of the current adult-based standards.

UNSAFE EXPOSURE

American children are routinely exposed to unsafe levels of OP insecticides in the food they eat. On any given day we estimate that more than one million children under age six exceed federal safety standards for OPs. One hundred thousand of these children exceed these same standards by a factor of 10 or more. The potential public health impact of these exposures is substantial, but as yet is not precisely understood.

For perspective, it is helpful to view the situation with OPs through the lens of experience with lead. For years lead was known to be toxic, but its special hazards to children, while suspected, were difficult to confirm. Only recently has science been able to bring into focus the subtle, yet profound learning deficits that result when infants and children are exposed to levels of lead that are perfectly safe for adults, and that were thought, until recently to be safe for children as well.

In some ways, the situation with OPs may be worse than lead because significant numbers of infants and children receive daily doses of multiple OPs that far exceed the safe dose for an adult. It is probable that these high OP exposures early in life are causing long term functional and learning deficits that scientists are just beginning to understand.

PESTICIDE-CONTAMINATED FOOD

Table 1. One out of every four times a child under 6 eats peaches, he or she is exposed to an unsafe dose of organophosphate insecticides (OPs).

Foods	Likelihood of being exposed to an unsafe dose of OPs
Peaches	24.8%
Apples	12.9%
Nectarines	12.2%
Popcorn	8.5%
Pears	7.5%
Cornbread	5.6%
Applesauce	5.2%
Grapes	5.1%
Corn Chips	4.5%
Pears (baby food)	3.8%
Raisins	3.3%
Cherries	3.2%
Kiwi	2.9%
Peaches (baby food)	2.4%
Apple Juice (baby food)	2.0%

Richard Wiles, Kert Davies, and Christopher Campbell, "Overexposed: Organophosphate Insecticides in Children's Food," January 1998.

Given this overwhelming evidence of unsafe exposure to organophosphate insecticides in the diet, EPA has little choice but to act to protect infants and children. The solution to the problem of unsafe levels of OPs in food, however, is not for children to eat less fruits and vegetables. Infants, children and pregnant women should be able to eat a diet rich in fruits and vegetables without any concern about short term illness or long term brain and nervous system damage that may result from unsafe levels of OP pesticides on these foods. The solution is to rid these healthful foods of the most toxic pesticides.

RECOMMENDATIONS

To begin to meet the requirements of FQPA and retain the greatest number of safe pesticides for farmers, several decisive but reasonable steps must be made. These actions would reduce risk from OPs to a level deemed acceptable under current EPA policy. We must emphasize again, however, that current EPA safety standards do not yet incorporate explicit or adequate protections for infants and children. Until reliable data on fetal and infant toxicity are available for all OPs, the actions recommended here, while significant, must be viewed as first steps in an ongoing process of protecting infants and children from OP insecticides.

First, all home and other structural use of OP insecticides must be banned. These uses put a small but significant number of infants and toddlers at extremely high risk, and in doing so jeopardize current agricultural uses of these compounds. Indeed, if food uses of any OPs are to be retained, all non-food uses with potential to expose pregnant women, infants or toddlers must be banned.

Second, at least five high risk OPs, methyl parathion, dimethoate, chlorpyrifos, pirimiphos methyl, and azinphos methyl, must be banned immediately for all agricultural use.

Third, all OPs must be banned for use in food that ends up in commercial baby food.

Fourth, EPA must require developmental neurotoxicity studies for all the remaining OPs found in the food supply. Prompt action can ensure that this critical information is available by the time EPA must take regulatory action on OPs in August 1999. At that time, the required additional ten-fold level of protection must be applied to any OP for which a developmental neurotoxicity study is not performed.

Fifth, food tolerances for all OPs must be lowered to levels that are safe for infants and children. To quote the National Research Council report, Pesticides in the Diets of Infants and Children, "Children should be able to eat a healthful diet containing legal residues without encroaching on safety margins." That is to say, legal residues, or tolerances, must be safe for infants and children. There is simply no scientific justification for retaining legal limits for pesticides in food that allow hugely unsafe levels of exposure, just because most children do not receive this exposure. This nonsensical notion is like leaving the speed limit at 500 miles per hour just because most people would still drive at 65.

| "Long-term health effects from the insecticide have been found only among people who were so sick that they nearly died—that is, among attempted suicides."

ORGANOPHOSPHATE INSECTICIDES SHOULD NOT BE BANNED

Michael Fumento

The Environmental Protection Agency must not ban the use of organophosphate insecticides, argues Michael Fumento in the following viewpoint. The insecticides are essential for killing and controlling pests indoors and on food crops, he maintains. Fumento also contends that studies have found that these insecticides have little effect on creatures other than insects. Furthermore, he asserts, prohibiting the use of these insecticides would be extremely costly to farmers and others who depend on them for eradicating insects. Fumento, a columnist for *Reason* magazine, is the author of *Polluted Science* and is a Senior Fellow at the Hudson Institute in Washington, D.C..

As you read, consider the following questions:

1. How do organophosphate insecticides kill insects, according to the author?
2. Why is the "additive" effect of pesticides a misguided theory, according to Marcia van Gemert as cited by Fumento?
3. What would be the result if the Environmental Protection Agency banned the use of organophosphate insecticides, in Fumento's opinion?

If you don't like bugs, 1998 could be a bad year. For by June 1998, the Environmental Protection Agency may promulgate the most sweeping anti-insecticide regulations in U.S. history. If it does, billions of dollars worth of crops may be lost annually, children may die from cockroach-related asthma and fire ant bites, and Lyme disease–carrying ticks may proliferate. And you may find that some of those raisins in your raisin bran, well, aren't.

At issue is a class of insecticides known as organophosphates. They kill bugs by interfering with their central nervous systems. Indoors they are a powerful remedy for cockroaches, fleas and termites. Outdoors they are used on practically every food crop you can name. For some crops there are absolutely no approved alternatives; for others the alternatives are either less effective or more expensive. In a 1994 study, the U.S. Department of Agriculture found that eliminating just one of the most common organophosphates, chlorpyrifos, would cost $150 million annually. According to Leonard Gianessi at the National Center for Food and Agricultural Policy in Washington, extrapolating this to a ban on all organophosphates—an option the EPA is seriously considering—would cost $1 to $2 billion a year.

POISONS ARE POISONOUS

The argument against organophosphates is essentially that they're poisonous—something that is true of most poisons. The question is how harmful they are to those of us with fewer than six legs. Answer: not very. Studies on laboratory mice have found that the average human adult would need to eat 875 pounds of broccoli every day for the rest of his life to approach the chlorpyrifos levels that caused problems in the rodents.

A 1997 EPA memorandum stated that chlorpyrifos "is one of the leading causes of acute insecticide poisoning incidents in the United States." That sounds ominous but isn't. Of almost 1,000 pesticides registered for use in the U.S., chlorpyrifos is the fourth most common. It's as if the government were claiming Fords are more dangerous than Ferraris because more Fords crash each year.

I spoke with numerous scientists and industry officials, and searched medical and newspaper databases, and I found not a single death or even near-death from chlorpyrifos that didn't result from intentional ingestion—that is, suicide or attempted suicide. True, some children have gotten hold of the chemical and, despite its bitter taste, become seriously ill from drinking it. But it appears all recovered fully. Long-term health effects from the insecticide have been found only among people who were so sick that they nearly died—that is, among attempted suicides.

Press accounts are rife with misinformation. The April 6, 1998, issue of *Newsweek*, for example, carries a small item called "The Pesticide Risk for Toddlers," which advises that toys be put away when chlorpyrifos spray "bombs" are used. What *Newsweek* doesn't mention is that these bombs are no longer distributed, and even if they were, there is no evidence that a child licking a toy sprayed with them would incur harm.

Still, the largest manufacturer of chlorpyrifos, DowAgrosciences of Indianapolis, has established a panel of experts to recommend ways to make its product safer. The company has joined other manufacturers and trade groups in a fight with government officials to make warning labels easier to read than current regulations allow. The company funds poison control centers across the country, and has given piles of money to fund more than 250 studies on organophosphates.

PREPOSTEROUS CLAIMS

That's not good enough to satisfy some environmentalists. Nothing short of a total ban on household use of organophosphates will suffice, the Environmental Working Group, a Washington-based nonprofit organization, asserted in a January 1998 paper. The report preposterously claims that each day a million children under five consume "unsafe levels" of organophosphates, primarily from residue on fruits. It demands, as a first step, an immediate ban on agricultural use of chlorpyrifos and several other organophosphates.

SAFE AND EFFECTIVE

Wormy apples, boll weevil–infested cotton, cockroaches in the kitchen—this is what the future could hold in America if the Environmental Protection Agency succeeds in banning organophosphates.

When applied as directed, these products are safe. We use them on our lawns and in our homes. They are one of the American farmer's most important lines of defense against insects.

Gene Ragan, *Dothan (Alabama) Eagle*, May 10, 1998.

"That would be foolish," says Barbara Petersen, a nutritional biochemist and head of Washington-based Novigen Sciences Inc. For 20 years she has evaluated pesticides and done dietary risk exposure studies under contract to both the EPA and industry, including some studies in which she collected grocery store samples. "We found extremely low levels" of chlorpyrifos, she says. "In the vast majority nothing was detected, and when we did detect residues, they were way below levels that EPA has set as permissible."

That's why the Environmental Working Group and other environmentalists—including top EPA officials—are now warning of the possible "additive" effect of pesticides, essentially arguing that harmless little bits can add up to a harmful level. Nonsense, says Marcia van Gemert, until recently head of the EPA Toxicology Branch. "You can't simply add two or three bodily risks together and conclude they cause a greater risk combined. Our [body] systems are a lot more complicated than that." Different chemicals, she adds, "have different targets, mechanisms, toxicities. They are really all quite different."

FAULTY SCIENCE

She also faults the Environmental Working Group's science, saying the organization "made a lot of mistakes in exposure estimates, and they have made 10 or 15 assumptions in which you couldn't retrace their steps." The attack on organophosphates, Ms. van Gemert adds, "is politically, not toxicologically, driven."...

One option suggested in an internal memo circulated by the EPA's Office of Pesticide Programs would "revoke all tolerances" for the chemicals. A tolerance is the amount of residue allowed on food, normally set at less than one-tenth the level that might harm anybody. Unless a crop has an established tolerance for a given pesticide, even one piece of fruit or vegetable in a shipment that has any detectable residue can cause the whole shipment to be seized and destroyed.

NEW POWERS

If the EPA were to revoke a pesticide's registration outright, farmers and chemical makers could mount an immediate legal challenge. But in the Food Quality Protection Act of 1996, Congress not only allowed the EPA to look at risks without considering benefits, but also expanded the agency's enforcement options. Among the agency's new powers is the ability to revoke tolerances. If the EPA follows this course, affected parties will have no way—legal or administrative—to introduce scientific studies, need or common sense into the equation.

Maybe the EPA will do the right thing. Maybe it won't drive fruit and vegetable prices up, ensuring that children eat less of them. Maybe it won't kill asthmatic children by banning potent roach-killing sprays. But a lot of little critters have their antennae crossed hoping otherwise.

PERIODICAL BIBLIOGRAPHY

The following articles have been selected to supplement the diverse views presented in this chapter. Addresses are provided for periodicals not indexed in the *Readers' Guide to Periodical Literature*, the *Alternative Press Index*, the *Social Sciences Index*, or the *Index to Legal Periodicals and Books*.

Michael Balter	"Thrust and Parry over Nuclear Risks," *Science*, January 31, 1997.
Sharon Begley	"Pesticides and Kids' Risks," *Newsweek*, June 1, 1998.
Sophie Boukhari	"Poison of the Earth," *Unesco Courier*, July/August 1998.
Betsy Carpenter	"Serving Up a Safer Food Supply," *U.S. News & World Report*, August 5, 1996.
Andrea Golaine Case	"Toxic Terror on the Golf Course?" *Priorities*, vol. 7, no. 2, 1995. Available from 1995 Broadway, 2nd fl., New York, NY 10023-5860.
Christopher H. Foreman Jr.	"And 'Environmental Justice' for All?" *Priorities*, vol. 9, no. 4, 1997.
Michael Fumento	"Pesticides Are Not the Main Problem," *New York Times*, June 30, 1998.
Linda Kulman	"6 Rms., Toxic Canal Vu," *U.S. News & World Report*, September 15, 1997.
Stephen Lester	"Children and Environmental Health," *Everyone's Backyard*, Summer 1997. Available from PO Box 6806, Falls Church, VA 22040.
Michelle Malkin and Michael Fumento	"The Mis-Measure of Man," *CEI Update*, April 1996. Available from 1001 Connecticut Ave. NW, Suite 1250, Washington, DC 20036.
Michael A. Pariza and Al Meyerhoff	"Should Pesticides That Cause Cancer in Animals Be Banned from Our Foods?" *Health*, March/April 1996.
Nancy Shute	"Is Your Lawn Making You Sick?" *McCall's*, April 1996.
Steve Stecklow	"New Food Quality Act Has Pesticide Makers Doing Human Testing," *Wall Street Journal*, September 28, 1998.
Workbook	Issue theme: Redefining Sludge, Summer 1998.

CHAPTER 3

IS RECYCLING AN EFFECTIVE RESPONSE TO POLLUTION?

Chapter Preface

During the 1980s, the question "Paper or plastic?" became a common phrase heard at many grocery stores as customers were given a choice between having their groceries placed in a paper sack or a plastic bag. For years, it was believed that the environmentally correct answer was "paper." Paper, after all, was biodegradable and could be recycled, while plastic filled up space in the landfill. But at the close of the twentieth century, the answer is not as clear as it once was.

Most paper is made from trees that are planted specifically to be made into paper and are thus a renewable resource. However, paper-making pollutes water supplies because it requires thousands of gallons of water to wash and bleach the wood pulp before it is made into paper. Once the paper is used and disposed of, it goes either to a landfill or to a recycling center. Modern landfills are lined with clay and plastic to protect the land and prevent contamination of groundwater. The garbage is covered with dirt daily, thus effectively eliminating any exposure to the elements that is necessary for the paper to biodegrade. Recycling paper consumes less energy and produces less air and water pollution than manufacturing paper from raw materials. However, paper recycling centers use hazardous chemicals to remove the ink from the paper and the resulting sludge must be cleaned and properly disposed of as well.

Producing and reusing plastic also has its advantages and disadvantages. Plastic is made from petroleum products, a nonrenewable resource. It is manufactured using electricity, much of which comes from nuclear power, which produces radioactive waste. Plastic is also nonbiodegradable; no matter how long it is in a landfill, it will never decompose. However, plastic takes up significantly less space than paper in a landfill by both weight and volume. Plastic can also be recycled, and the energy used to recycle plastic is less than the energy required to produce virgin plastic.

Most analysts agree that the benefits and costs of using paper or plastic are nearly equal, leaving the decision of which to use to the consumer's preference. In the following chapter, the authors examine whether recycling's advantages outweigh its costs and whether recycling can reduce pollution.

> "In virtually all cases, recycling helps reduce or eliminate the pollution typically associated with the production and disposal of consumer products."

RECYCLING CONSERVES THE ENVIRONMENT

Allen Hershkowitz

In the following viewpoint, Allen Hershkowitz argues that recycling and using recycled materials in manufacturing processes produce less pollution than using raw materials. Recycling creates fewer air and water pollutants, he contends, and consumes less energy than manufacturing aluminum, glass, paper, and some plastics from virgin materials. Hershkowitz is a senior scientist with the Natural Resources Defense Council, which published the report "Too Good to Throw Away," from which this viewpoint is excerpted.

As you read, consider the following questions:

1. According to Hershkowitz, why do paper recycling mills emit less pollution than virgin paper mills?
2. What evidence does the author present to support his contention that garbage trucks cause more pollution than recycling trucks?
3. What percentage of paper made in the United States is produced from trees planted on tree plantations?

Excerpted from Allen Hershkowitz, *Too Good to Throw Away: Recycling's Proven Record*, Natural Resources Defense Council, February 1997. Reprinted by permission of the Natural Resources Defense Council. Endnotes in the original have been omitted in this reprint.

U sing recycled materials helps avoid the air and water pollution typically caused by manufacturing plants that rely
solely on unprocessed, virgin raw materials. Because using recycled materials reduces the need to extract, process, refine, and
transport timber, crude petroleum, ores, etc., into virgin-based
paper, plastics, glass, and metals, recycling lessens the toxic air
emissions, effluents, and solid wastes that these manufacturing
processes create. It is virtually beyond dispute that manufacturing products from recyclables instead of from virgin raw materials—making, for instance, paper out of old newspapers instead
of virgin timber—causes less pollution and imposes fewer burdens on the earth's natural habitat and biodiversity.

THE CHARGES

Antienvironmental theorists dismiss these benefits. The Cato Institute, a conservative research and advocacy group based in
Washington, D.C., claims that state and local ordinances that
promote recycling "neither conserve scarce resources nor help
to protect the environment." According to the Reason Foundation, "Recycling itself can cause environmental harm." Most recently, the benefits recycling provides in avoiding pollution
caused at manufacturing plants were dismissed by John Tierney
in "Recycling Is Garbage" in the New York Times Magazine as follows:

> [T]here are much more direct—and cheaper—ways to reduce
> pollution. Recycling is a waste of . . . natural resources . . . [and]
> a messy way to try to help the environment.

The Facts In virtually all cases, recycling helps reduce or eliminate the pollution typically associated with the production and
disposal of consumer products. Antirecycling interests who argue
otherwise are either out of touch with or conveniently ignoring
well known and widely documented environmental facts. . . .

THE RECYCLING DIFFERENCE

The most controversial charge leveled by the antirecycling crowd
is that recycling really does not benefit the environment; that, to
the contrary, it produces its own significant pollution. For example, the Cato Institute has claimed that "de-inking 100 tons of
old newspaper for subsequent reuse generates 40 tons of toxic
waste." According to the Reason Foundation, "The environmental costs of recycling may exceed any possible environmental
benefits." Here's how John Tierney put it in the Times Magazine:

> [R]ecycling operations create pollution in areas where more
> people are affected: fumes and noise from collection trucks,

solid waste and sludge from the mills that remove ink and turn the paper into pulp. Recycling newsprint actually creates more water pollution than making new paper from virgin sources. . . . For each ton of newsprint that's produced, an extra 5,000 gallons of waste water are discharged.

The Facts Less water pollution is produced per ton by paper recycling mills than by virgin paper mills. The recycled sector of the global paper industry, being more recently developed, is in fact the industry's most modern, efficient, and least polluting sector. The charge made by the Cato Institute that newsprint recycling mills produce forty tons of "toxic" waste for every hundred tons of paper recycled is, simply, absurd. Far from producing more hazardous pollution than virgin mills, modern paper recycling mills produce virtually no hazardous air or water pollution or hazardous wastes. Even the most cursory review of engineering designs for newsprint recycling mills reveals that the product yield per ton of recovered paper used by the mill is in the range of 80 to 85 percent—in other words, only about fifteen to twenty tons end up as waste per hundred tons of manufactured paper. None of this waste is "toxic." For virgin mills, the ratio is nearly the opposite: 75 percent of the harvested tree does not wind up as paper product. Where would the "toxic" residues allegedly produced by paper recycling mills come from? Less than one percent of the waste from a recycled paper mill is from ink, which is today more properly described as benign vegetable dye or carbon coated with plastic polymers; the remaining waste is water (90 percent) and short paper fibers (about 10 percent). No recycled paper mill could operate successfully, financially or otherwise, with a 60 percent product yield, as the Cato Institute has claimed, especially if the by-product generated was "toxic waste," which requires special, extremely costly handling, treatment, and disposal according to U.S. federal hazardous waste laws.

PAPER MADE FROM RECYCLABLES

Contrary to the claim that the recycling of newspaper produces an extra "5,000 gallons of wastewater" per ton of newsprint manufactured, new paper mills that recycle 100 percent newsprint do not even consume or discharge a total of 5,000 gallons of water per ton of manufactured product. By contrast, according to the Federal Register, volume 58, number 241, the virgin "pulp and paper industry is the largest industrial process water user in the United States." Virgin newsprint mills, at best, use about the same amount of water as recycled newsprint mills, but in fact most virgin newsprint mills use more water per ton of manufactured product, sometimes twice as much. For example: A 100 percent recycled newsprint mill in Aylesford, England, uses and discharges less than 4,000 gallons of water per ton of manufactured product. A 100 percent recycled newsprint mill under development in New York City's South Bronx will consume about 3,800 gallons of water per ton of manufactured product, and more than 80 percent of that water will come from a sewage treatment plant as recovered and cleaned effluent. Another mill under construction on Staten Island in New York City, which will recycle newspapers and other types of wastepaper, will also use less than 4,000 gallons of water for every ton of manufactured product. By contrast, virgin newsprint mills that are ten years old or older use approximately 10,000 gallons of water per ton of manufactured newsprint, while the few modern mills that have been built in the past eight to ten years use 4,000 to 5,000 gallons of water per ton of manufactured newsprint. And most newsprint purchased in the United States is produced at older Canadian mills. Overall, the recycled paper industry is proving to be the least polluting and the most modern, efficient sector in the paper manufacturing industry. . . .

TRANSPORTING RECYCLABLES

It is also wrong to argue that pollution from recycling vehicles in urban areas is greater than pollution from trucks and infrastructure dedicated to garbage collection and disposal. At worst, recycling trucks and infrastructure may produce the same amount of pollution. But, as explained below, it is more likely that diverting waste to recycling processing facilities tends to cause less pollution than garbage collection does.

Any added pollution that might be generated by a recycling program would result from additional trucks dedicated to collection. However, because in most cities "the fleet of trucks to collect recyclables is substantially smaller and less costly than the [typical] waste [collection] fleet," write Peter Anderson, George Dreckman, and John Reindl, the types of vehicles used to collect recyclables typically generate less pollution than standard waste collection vehicles. Moreover, collecting recyclables has been shown to be a faster operation than collecting garbage. They add, "It takes less time at the curb to pick up a load of recyclables, which is typically under 10–15 pounds, than solid waste, which can be as much as 50 pounds or more." This means a truck collecting recyclables, on average, idles for a shorter time than does a garbage truck, thus emitting less pollutants. Finally, by diverting waste into smaller, less polluting recycling vehicles, real-world experience confirms that there can be a reduction in the number of larger trucks and, indeed, of entire truck routes dedicated to garbage collection. . . . The higher the recycling rate, the fewer trucks needed for garbage collection. Thus, *the strategy to reduce any additional pollution that duplicate trucks might cause is to increase recycling.* In that way trucks dedicated to garbage collection can be retired.

Moreover, because landfills are usually located far from the concentrated population centers that generate municipal wastes, transporting garbage to landfills typically requires more vehicle miles to be traveled than does the deposit of recyclables at processing plants, which are usually numerous and more frequently located within city limits. In New York City, more than two dozen recycling processing plants conveniently located in every borough help reduce the vehicle miles traveled to deposit recyclables. By contrast, the non-recyclable portion of the city's waste is deposited at any one of only eight marine transfer stations by the city's fleet of 2,500 collection vehicles, or directly at the logistically remote Fresh Kills landfill on Staten Island. Were New York City to adopt a policy of long-hauling waste for export, as dozens of other cities do, even more vehicle miles would be traveled, increasing mobile-source pollution.

SAVING TREES

One of the very first environmental lessons most children learn is that recycling paper saves trees. That critical fact has been a launching pad for turning millions of families into environmentally concerned citizens. Antienvironmental interests have attempted to remove that attractive motivating image from the recycling lexicon. In its attempt to do so, the Reason Foundation has claimed "most of the trees used to make paper are not from virgin forests, but trees planted explicitly for manufacturing paper." According to John Tierney's antirecycling article in the *Times Magazine*:

> Yes, a lot of trees have been cut down to make today's newspaper. But even more trees will probably be planted in their place. America's timber supply has been increasing for decades. . . . Paper is an agricultural product, made from trees grown specifically for paper production. Acting to conserve trees by recycling paper is like acting to conserve cornstalks by cutting back on corn consumption.

The Facts It is specious to say that "virgin forests" are not used to make paper since virtually all the virgin forests in the United States have already been cut down. As the Organization of Eco-

ENERGY SAVINGS AND CO_2 IMPACTS: RECYCLING AND INCINERATION

Material	Grade	Energy Savings Per Ton Recycled			Energy Generated Per Ton Incinerated
		% Reduction of Energy*	Million BTUs	Tons CO_2 Reduced	Million BTUs
Aluminum		95	196	13.8	−1.06
Paper**	Newsprint	45	20.9	−0.03	11.8
	Print/Writing	35	20.8	−0.03	11.8
	Linerboard	26	12.3	0.07	11.8
	Boxboard	26	12.8	0.04	11.8
Glass	Recycle	31	4.74	0.39	−0.34
	Reuse	328	50.18	3.46	na
Steel		61	14.3	1.52	−0.34
Plastic	PET	57	57.9	0.985	35.9
	PE	75	56.7	0.346	35.9
	PP	74	53.6	1.32	38.5

* Relative to energy required for virgin production
** Energy calculations for paper recycling count unused wood as fuel
na = not applicable

Allen Hershkowitz, "Too Good to Throw Away: Recycling's Proven Record," February 1997.

nomic Cooperation and Development (OECD) has confirmed, throughout the United States, "95 per cent of the originally forested [virgin] area has been cleared and logged." Moreover, the claim that "most of the trees used to make paper are not from virgin forests, but trees planted explicitly for manufacturing paper" is inaccurate. In fact, only a small percentage—less than 20 percent—of paper manufactured in the United States comes from tree plantations. Yes, a lot of trees are cut down to make newspaper, and the types of trees used to manufacture newspaper are not being replaced at the rate at which they are harvested. A bevy of recently issued timber industry analyses confirm that "[u]nfavorable growth/drain ratios clearly suggest that we are depleting our 'forest capital.'" Virtually all professional analysts of wood fiber supply in the United States report enormous pressure on commercially available timber. . . .

THE ENERGY BONUS

Recycling is the most energy conserving of all waste management strategies. Numerous industry and government studies have repeatedly documented that the collection and use of secondary materials results in large energy savings over traditional production and disposal methods. Net energy savings may vary from product to product and region to region, as well as from production facility to facility, but there is no doubt that energy use reductions are realized across the board by recycling. There is not much debate about this, and even antirecycling interests generally concede the point. According to John Tierney, "recycling does at least save energy—the extra fuel burned while picking up recyclables is more than offset by the energy savings from manufacturing less virgin paper, glass and metal." The *Wall Street Journal* concurs: "To be sure, reuse of old paper, metals, glass and even some plastics makes great sense. It almost always lowers raw-materials costs in manufacturing, usually reduces energy consumption and in some cases cuts air and water pollution.". . .

Meeting the needs of the present generation without compromising the ability of future generations to meet their own needs is the fundamental principle underlying the concept of sustainability. In both a material sense and in the way it fosters community participation and a concern for unseen people, faraway places, and future generations, few policies advance sustainability as much as recycling. The antirecycling message is being widely disseminated and, invariably, it is generating pro-environment responses from average citizens as well as representatives from all levels of government. Far from trashing recycling and impugning the motives of its proponents, all sectors of the polity would do well, materially and spiritually, to embrace and help advance the sustainable, community-building, natural harmony it promotes.

"There is no environmental reason to recycle trash."

RECYCLING DOES NOT CONSERVE THE ENVIRONMENT

Doug Bandow

In the following viewpoint, Doug Bandow contends that most trash is not worth the time or money spent to recycle it. Recycling programs frequently lose money because it costs more to transform recycled materials into products than it does to produce the item using virgin materials, he asserts. Moreover, Bandow concludes, recycling trash and newspapers usually generates more pollution and consumes more energy than it saves. Bandow is a senior fellow at the Cato Institute, a libertarian public policy think tank in Washington, D.C.

As you read, consider the following questions:

1. How does putting packaging on food products reduce waste, according to Tierney as cited by Bandow?
2. How big of a landfill would be necessary to contain all the expected trash generated by Americans over the next 1,000 years, in A. Clark Wiseman's opinion?
3. Who benefits the most from recycling, in Bandow's view?

Reprinted from Doug Bandow, "Our Widespread Faith in Recycling Is Misplaced," *This Just In*, August 27, 1997, by permission of the Cato Institute.

T he Earth. It's hard not to like it. Many people adore it. Indeed, there has long been a strand of environmentalism that
treats nature as divine. So-called Deep Ecologists, for instance,
term their "eco-terrorist" attacks acts of worship to the planet.
Few Americans would go so far, of course, but many of them
worship in their own way. They recycle.

A decade ago a wandering garbage barge set off a political crisis: Where will we put our trash? The media inflamed people's
fears of mounting piles of garbage. A variety of interest groups—
particularly "public relations consultants, environmental organizations, waste-handling corporations," according to journalist
John Tierney—lobbied to line their pockets. Politicians seeking
to win votes enacted a spate of laws and regulations to encourage
and often mandate recycling.

MORAL FERVOR

But while politics did help create an industry, it did not generate
the moral fervor behind it. Many people see recycling as their
way to help preserve the planet. For some, it may be the environmental equivalent of serving time in Purgatory, attempting
to atone for the materialist excesses of a consumer society. It allows one to feel good about oneself even while enjoying every
modern convenience.

This moral fervor comes at a price. A study from the Reason
Foundation, "Packaging, Recycling, and Solid Waste," concludes
that recycling, though sometimes beneficial, all too often wastes
resources. But then, it has long been known that most trash isn't
worth reusing, recycling programs usually lose money, and
landfills offer a safe disposal method.

Indeed, in 1996 John Tierney wrote a devastating article for
the New York Times Magazine titled "Recycling is Garbage." He declared that the emperor had no clothes: "Recycling may be the
most wasteful activity in modern America: a waste of time and
money, a waste of human and natural resources."

His points were many. For instance, packaging saves resources, reducing food spoilage. Fast-food meals generate less
trash per person than do home-cooked meals. The cheapest way
to dispose of garbage is in a landfill. Modern dumps incorporate
a range of safeguards and take up a minuscule amount of space.

A. Clark Wiseman of Spokane's Gonzaga University figures
that, at the current rate, Americans could put all of the trash
generated over the next 1,000 years into a landfill 100 yards
high and 35 miles square. Or dig a similar-size hole and plant
grass on top after it was filled.

Recycling, in contrast, costs money. New York City's mandatory program spends $200 more per ton to collect recyclables than it would cost to bury them, and another $40 per ton to pay a company to process them. Tierney figures the value of the private labor wasted complying with the rules (rinsing, taking off labels, sorting) to be literally hundreds of dollars more per ton.

A Trivial Pursuit

Consumers are most deceived about plastics recycling, especially plastic packaging. They are convinced that vast amounts of fossil fuels will be wasted unless plastics are collected and reprocessed.

But plastics use less than 3 percent of all the energy consumed in the United States annually. Plastic products use 7 percent of the natural gas and 2 percent of the oil consumed. Plastic packaging is a fraction of the total plastic end uses and consumes less than 1 percent of all fossil fuels used in the United States. Thus, even if plastic packaging could be recycled at a 25 percent rate, the net saving in fossil fuel use clearly is trivial pursuit—0.25 percent of the total.

Kenneth W. Chilton, Mackinac Center for Public Policy's *Viewpoint on Public Policy*, January 12, 1998.

Yet there is no environmental reason to recycle trash. Resources are not scarce. In fact, much newsprint comes from trees grown for that specific purpose. Even Worldwatch, a reliably hysterical group that has constantly (though luckily, so far inaccurately) predicted impending environmental doom, now acknowledges: "The question of scarcity may never have been the most important one."

Recycling Is Wasteful

Moreover, making recyclables generates waste. For instance, producing paperboard burger containers yields more air and water pollution and consumes more energy than does manufacturing polystyrene clamshells. It takes more water to recycle newsprint than to make it afresh.

How can such a wasteful practice persist? Tierney concluded: "By turning garbage into a political issue, environmentalists have created jobs for themselves as lawyers, lobbyists, researchers, educators and moral guardians. Environmentalists may genuinely believe they're helping the Earth, but they have been hurting the common good while profiting personally."

Tierney's article infuriated environmentalists, but was ignored by business, which is paying much of the cost of the re-cycling liturgy. Only silence emanated from companies that have the most to gain from returning garbage to the marketplace.

Yet inaction is a prescription for more regulation. The federal government is considering increasing its national objective for recycling from 25 percent to 35 percent, 41 states already impose some form of goal or mandate regarding trash reduction and recycling, and some jurisdictions are considering new laws, such as so-called advance disposal fees. Politicians who care little about facts and feel political pressure only from environmentalists are likely to add new rules and toughen existing ones.

If people want to worship the Earth by recycling, they are certainly free to do so. But the government shouldn't dragoon skeptics into the same wasteful ceremonies. It is time for an environmental reformation, in which lawmakers change public policy to reflect the wastefulness of recycling.

"When designed right, recycling programs are cost-competitive with trash collection and disposal."

RECYCLING IS ECONOMICAL

Brenda Platt

Brenda Platt is the director of Materials Recovery for the Institute for Local Self-Reliance, an education and research organization on environmental and economic policies. In the following viewpoint, Platt contends that many arguments used against recycling are based on ill-founded myths. Recycling programs can be cost-competitive with trash collection, landfills, and incinerators, she argues. Furthermore, recycling programs create more jobs than landfills and incinerators, Platt asserts. Intervention by the public sector in waste disposal is needed only to ensure that complex laws and regulations concerning waste disposal are not biased against recycling.

As you read, consider the following questions:
1. What percentage of the waste stream could be recycled, according to studies cited by the author?
2. What should be considered when determining whether recycling and composting is cost-competitive with conventional means of trash collection and disposal, according to Platt?
3. In the author's opinion, how do the current rules concerning trash disposal favor landfills and incinerators over recycling and other disposal alternatives?

Reprinted from Brenda Platt, "The Five Most Dangerous Myths About Recycling," September 1996 publication on the Institute for Local Self-Reliance website: www.ilsr.org/recycling/fivemyths.html. Used with permission. Endnotes in the original have been omitted in this reprint.

M yth #1:*We can recycle only 25 to 30% of our solid wastes.*
    Fact: Twenty-five percent was considered a maximum
level in 1985. Today it should be considered a minimum, not a
maximum. By continuing to build the reuse, recycling, and
composting infrastructure and integrating the best features from
the best programs—local and state—the nation as a whole can
achieve 50% recycling by 2005.

Recycling continues to grow. Between 1980 and 1990, the
United States almost doubled its recycling rate from 9% to 17%.
By 1995, the nation's recycling rate had jumped to 27%. The
growth in recycling and composting programs contributed to
this dramatic increase. The number of curbside programs grew
from 1,000 in 1988 to 7,375 in 1995. During the same period,
the number of yard trimmings composting operations increased
from 700 to 3,316. A dozen states are recycling 30% or more of
their municipal solid wastes—Delaware, Florida, Maine, Massachusetts, Minnesota, New Jersey, New York, North Carolina,
Ohio, Oregon, Virginia, and Washington. Within these states,
hundreds of communities have reached 50% and higher levels.
These record-setters are implementing waste prevention strategies, targeting a wide range of materials for recovery, offering
convenient service (curbside and drop-off collection are both
important), employing collection and processing techniques that
encourage resident participation as well as yield high-quality
materials, establishing strong economic incentives—particularly
volume-based trash rates, collecting source-separated yard waste
for composting, encouraging backyard composting, and extend-
ing programs beyond the residential sector to the commercial
and institutional sectors. Programs around the country continue
to expand and improve, serving more people and targeting new
materials for recovery. The recovery of food discards, textiles,
construction and demolition materials, and reusable items, for
instance, are all on the rise. Several studies indicate that 60 to
80% of the waste stream is indeed recoverable.

THE ECONOMICS OF RECYCLING

Myth #2: *Recycling is more expensive than trash collection and disposal.*

Fact: When designed right, recycling programs are cost-competitive with trash collection and disposal.

The cost of curbside recycling is often compared with the
cost of conventional disposal alone, even though the cost of recycling displaces collection as well as disposal costs. The average
cost of collection and disposal should be compared to the overall average cost of collection and recovery. When this compari-

son is made, the economics of recycling and composting often look very impressive. Data from a nationwide survey of 264 recycling programs suggest that recycling is cost-effective once landfill tip fees reach $33 per ton. Many recycling and composting programs remain cost-effective at much lower fees. The survey, for instance, found that mandatory programs—which have lower per ton costs as a result of higher participation and higher amounts of materials collected—are cost competitive with landfill tipping fees of $15 per ton. (Average regional landfill fees range from $65 per ton in the Northeast to $16 per ton in the Rocky Mountain region.) Other studies have also concluded that recycling costs less than traditional trash collection and disposal when communities achieve high levels of recycling. It is true that in some communities recycling is expensive. But often that is because these communities are still recycling at very low rates and are treating recycling as an add-on to their traditional trash system rather than as a replacement for it. Communities that maximize recycling save money by redesigning their collection schedules and/or trucks.

THE MOST PRODUCTIVE ACTIVITY IN AMERICA

The case for recycling is strong. The bottom line is clear. Recycling requires a trivial amount of our time. Recycling saves money and reduces pollution. Recycling creates more jobs than landfilling or incineration. And a largely ignored but very important consideration, recycling reduces our need to dump our garbage in someone else's backyard.

Come to think of it, recycling just might be the most productive activity in modern America.

David Morris, St. Paul Pioneer Press, July 30, 1996.

Recycling critics erroneously assume that virtually all the costs of the solid waste system are fixed, that is, represent long-term capital investments. This assumption leads them to view recycling as an add-on cost and therefore expensive. In fact, when recycling reaches high levels and system managers view it as the way they collect wastes, then fixed costs can become variable costs. Labor can be reassigned. Twenty percent of the fleet vehicles turn over annually and can be redesigned and reduced in scale and cost. Baltimore, Maryland, for example, uses the same conventional trash trucks to collect recyclables and trash, separately and at different times. This minimized its upfront costs and allowed Baltimore to add recycling with no increase in its solid waste budget. Loveland, Colorado, uses the same vehicles to collect recyclables and trash, but does so simultaneously. Loveland recovers 56% of its residential waste. Cost per household did not rise when the City added recycling. Plano, Texas replaced one of its two trash collection days with collection of recyclables and yard waste at no additional costs. Takoma Park, Maryland did the same. The City avoided hiring additional employees by splitting collection crews between recycling and trash. Not only has the number of trucks remained the same, but they have not been replaced and need less maintenance as a result of decreased trash collected; half of Takoma Park's waste is recovered. As communities attain ever higher recovery levels, planners and public works administrators are beginning to realize that recycling and composting can be the primary strategy for handling our solid wastes, rather than a supplement to the conventional system. The economics of materials recovery improves when, instead of adding the costs of recycling and composting onto the costs of conventional collection and disposal, the two are integrated.

RECYCLING IS COST-EFFECTIVE

Myth #3: Landfills and incinerators are more cost-effective and environmentally sound than recycling options.

Fact: Recycling programs, when designed properly, are cost-competitive with landfills and incinerators, and provide net pollution prevention benefits. Recycling materials not only avoids the pollution that would be generated through landfilling and incinerating these, but also reduces the environmental burden of virgin materials extraction and manufacturing processes.

Even when landfill tipping fees are low, recycling and composting may still be preferable to disposal options. At least 22 states have less than 10 years of landfill capacity left. Southern states reportedly average five years of remaining capacity. New landfills may cost far more than existing ones. Recent U.S. Environmental Protection Agency (EPA) rules requiring municipal landfills to install liners and leachate collection systems are closing hundreds of landfills. One result is the trend toward fewer but larger and privately owned landfills. Fewer landfills will mean increased transportation costs. There may be no scarcity of land for new landfills but new landfills being built tend to be quite remote from population centers. Long hauling and disposing municipal solid waste at distant landfills is already costing some cities on the West and East Coasts between $40 and $70 per ton. Privately owned landfills may increase costs. One study found that publicly owned landfills are 20% less expensive than privately owned landfills and provide greater local control over disposal activities. Thus, existing landfills are a precious possession. Recycling extends their lives. Projected as well as current costs and availability of landfills should be taken into account in any evaluation of the cost-effectiveness of waste reduction and recovery options. In addition, while we may have no shortage of nearby land to dump our trash, few communities want to be dumped upon. Thus, national policy should be to reduce the burden on the environment and on local communities from the transportation and dumping of trash.

Many communities have turned to incineration as an alternative to landfills. But incinerators are expensive. Tip fees at incinerators built between 1989 and 1993 average $60 per ton. Some more recent incinerators have had to lower tip fees in order to compete with other disposal facilities. Montgomery County, Maryland, for instance, increased taxes to property owners to cover the operating costs of its newly built incinerator after it lowered the facility's tip fees in order to attract waste. Incinerators are always the most capital-intensive solid waste management option; materials recovery can be the least. While landfills pollute—one out of every five Superfund toxic waste sites is a former municipal solid waste landfill, and, even the best landfills will eventually leak, contaminating groundwater—incinerators are potentially more polluting. Thirty percent by weight of trash entering incinerators exits as ash, a waste product that may contain high levels of toxic residues. Moreover, incinerators emit organic compounds, carbon dioxide, sulfur dioxide, nitrogen oxide, and other acid gases that landfills do not. Incineration has another drawback—it competes with recycling and composting programs for the same materials. A study evaluating Florida's seven largest incinerators found that these facilities regularly burn significant amounts of highly recyclable materials. "Put-or-pay" contracts, which require local governments to deliver guaranteed tonnage of waste to incinerators, are a major disincentive to maximizing recycling or waste reduction, and thus an obstacle to low-cost materials recovery programs.

RECYCLING CREATES JOBS

Myth #4: *Landfills are significant job generators for rural communities.*

Fact: Recycling creates many more jobs for rural and urban communities than landfill and incineration disposal options.

Just sorting collected recyclable materials sustains, on a per-ton basis, 10 times more jobs than landfilling. However, it is making new products from the old that offers the largest economic pay-off. New recycling-based manufacturers employ even more people and at higher wages. Recycling-based paper mills and plastic product manufacturers, for instance, employ 60 times more workers than do landfills. Product reuse also sustains significantly more jobs than disposal options. Computer refurbishing and repair, for example, creates 68 times more jobs than landfills. If half the 25.5 million tons of durable goods now discarded into America's landfills each year were reclaimed through reuse, more than 100,000 new jobs could be created in this industry alone.

PUBLIC-SECTOR INTERVENTION

Myth #5: The marketplace works best in solving solid waste management problems; no public-sector intervention is needed.

Fact: The solid waste system has always operated under public sector rules and always will. Currently these rules encourage unchecked product consumption and disposal. Public-sector intervention is needed to shape a system in which materials are produced, used, discarded, and recovered efficiently. We need to change the rules so that disposal alternatives—source reduction, reuse, recycling, and composting—operate in a level playing field. Even after we level the playing field, favoring disposal alternatives makes sense because of its many community and public sector benefits.

Our solid waste management systems do not flow naturally from a preordained plan or even from the free market. They are governed by a complex set of rules and regulations, international agreements to local ordinances and everything in-between. These rules take many forms—tax laws, virgin materials subsidies, business regulations, environmental laws, land use requirements, the commerce clause, flow control, the Public Utilities and Regulatory Act—but together they help shape what sort of waste management infrastructure thrives. Right now, these rules favor a one-way flow of materials from the producer, to the consumer, to the dump or incinerator, and a system in which trash collection and disposal is falsely viewed as cost-effective while more efficient materials use through source reduction, reuse, recycling, and composting is falsely viewed as having to pay for itself. Seventy-three percent of our municipal solid waste ends up in landfills or incinerators. We need to establish rules that will instead fashion a system in which materials are produced and utilized efficiently with minimal environmental impacts and maximum sustainable economic development benefits.

Solid waste, by definition, represents inefficiencies, and moreover, no one wants a landfill or incinerator in their backyard. Thus, favoring disposal alternatives through public-sector intervention makes sense even when a level playing field exists with disposal. Source reduction, reuse, recycling, and composting extend landfills, reduce our need to dump our trash in someone else's backyard, reduce pollution, make us more frugal, and create jobs and new businesses. Communities should have the right to take these qualitative, quality of life elements into account when they make the rules. Many of those states with the highest recycling rates have met with success because they have begun to change the rules under which the marketplace operates by enacting bottle bills, minimum recycled-content product standards, landfill disposal bans, mandatory recycling, recycling-related business attraction incentives for the private and community sectors, and procurement programs for recycled-content products. These public-sector interventions are needed in order to transform solid waste management problems into materials conservation and recovery opportunities.

"Collecting a ton of recyclable items is three times more expensive than collecting a ton of garbage."

RECYCLING IS UNECONOMICAL

John Tierney

John Tierney argues in the following viewpoint that recycling waste such as glass, plastic, and paper costs more than it saves in time, money, energy, and pollution. Costs associated with recycling exceed what cities can make on selling the recyclable materials, he maintains. Furthermore, Tierney asserts, there is no shortage of landfill space, so disposing of waste in a landfill is usually cheaper than recycling it in most parts of the country. Moreover, many recycling practices that are meant to help the environment are actually harmful, he contends. Tierney is a staff writer for the *New York Times Magazine*.

As you read, consider the following questions:

1. What are some of the factors that must be considered when determining how much recycling costs a city, in Tierney's opinion?
2. What is recycling's only benefit, according to researchers cited by the author?
3. According to Tierney, what is the simplest and best measure of a product's environmental impact?

The 1992 Plan projected that the City would realize net savings from recycling. The Department's experience to date in implementing the recycling program diverges from the assumptions of the Plan.

—1996 Comprehensive Solid Waste Management
Plan of the New York City Department of Sanitation.

Every time a sanitation department crew picks up a load of bottles and cans from the curb, New York City loses money. The recycling program consumes resources. It requires extra administrators and a continual public relations campaign explaining what to do with dozens of different products—recycle milk jugs but not milk cartons, index cards but not construction paper. (Most New Yorkers still don't know the rules.) It requires enforcement agents to inspect garbage and issue tickets. Most of all, it requires extra collection crews and trucks. Collecting a ton of recyclable items is three times more expensive than collecting a ton of garbage because the crews pick up less material at each stop. For every ton of glass, plastic and metal that the truck delivers to a private recycler, the city currently spends $200 more than it would spend to bury the material in a landfill.

Officials hoped to recover this extra cost by selling the material, but the market price of a ton has never been anywhere near $200. In fact, it has rarely risen as high as zero. Private recyclers usually demand a fee because their processing costs exceed the eventual sales price of the recycled materials. So the city, having already lost $200 collecting the ton of material, typically has to pay another $40 to get rid of it.

THE LABOR COST

The recycling program has been costing $50 million to $100 million annually, and that's just the money coming directly out of the municipal budget. There's also the labor involved: the garbage-sorting that millions of New Yorkers do at home every week. How much would the city have to spend if it couldn't rely on forced labor? True, some people would probably be glad to do the work for free because they regard garbage-sorting as a morally uplifting activity for the whole family. But many others have refused to follow the law. They seem to have a more traditional view of garbage-sorting: an activity done only for money, and then only by the most destitute members of society.

I tried to estimate the value of New Yorkers' garbage-sorting by financing an experiment by a neutral observer (a Columbia University student with no strong feelings about recycling). He kept a record of the work he did during one week complying with New York's recycling laws. It took him eight minutes during the week to sort, rinse and deliver four pounds of cans and bottles to the basement of his building. If the city paid for that work at a typical janitorial wage ($12 per hour), it would pay $792 in home labor costs for each ton of cans and bottles collected. And what about the extra space occupied by that recycling receptacle in the kitchen? It must take up at least a square foot, which in New York costs at least $4 a week to rent. If the city had to pay for this space, the cost per ton of recyclables would be about $2,000. That figure plus the home labor costs, added to what the city already spends on its collection program, totals more than $3,000 for a ton of scrap metal, glass and plastic. For that price, you could find a one-ton collection of those materials at a used-car lot—a Toyota Tercel, for instance—and drive home in it.

A MONEY-LOSING PROPOSITION

In 1995, a surge in the market price for recycled materials prompted a spate of recycling-has-finally-arrived articles. At one point, New York was selling its old newspapers for $150 per ton, which was almost enough to offset the extra costs of the paper recycling program. But newsprint prices have since plummeted back to familiar levels; New York is once again paying recyclers to take its newspapers, and city officials are resigned to losing money on recycling. As a result of a lawsuit by City Council members and the Natural Resources Defense Council, the city has been under court order to collect increasing amounts of recyclable material to meet goals set in law. City officials have promised to comply by expanding the recycling program and promoting a separate program in the public schools, but they've been stalling because they don't want to increase the budget deficit.

Officials in some cities claim that curbside recycling programs are cheaper than burying the garbage in a landfill, which can be true in places where the landfill fees are high and the collection costs aren't as exorbitant as in New York. But officials who claim that recycling programs save money often don't fully account for the costs. "A lot of programs, especially in the early years, have used funny-money economics to justify recycling," says Chaz Miller, a contributing editor for Recycling Times, a trade newspaper. "There's been a messianic zeal that's hurt the cause. The American public loves recycling, but we have to do it efficiently. It should be a business, not a religion."

Recycling programs didn't fare well in a federally financed study conducted by the the Solid Waste Association of North America, a trade association for municipal waste-management officials. The study painstakingly analyzed costs in six communities (Minneapolis; Palm Beach, Fla.; Seattle; Scottsdale, Ariz.; Sevierville, Tenn., and Springfield, Mass.). It found that all but one of the curbside recycling programs, and all the composting operations and waste-to-energy incinerators, increased the cost of waste disposal. (The exception was Seattle's curbside program, which was slightly cheaper—by one-tenth of 1 percent—than putting the garbage in a landfill.) Studies in European cities have reached similar conclusions. Recycling has been notoriously unprofitable in Germany, whose national program is even less efficient than New York's.

"We have to recognize that recycling costs money," says William Franklin, an engineer who has conducted a national study of recycling costs for the not-for-profit group Keep America Beautiful. He estimates that, at today's prices, a curbside recycling program typically adds 15 percent to the costs of waste disposal—and more if communities get too ambitious.

COST-BENEFIT ANALYSIS

Franklin and other researchers have concluded that recycling does at least save energy—the extra fuel burned while picking up recyclables is more than offset by the energy savings from manufacturing less virgin paper, glass and metal. "The net result of recycling is lower energy consumption and lower releases of air and water pollutants," says Richard Denison, a senior scientist at the Environmental Defense Fund, which has calculated the ecological benefits of recycling. But there are much more direct—and cheaper—ways to reduce pollution. Recycling is a messy way to try to help the environment. Consider a few questions whose answers would seem obvious to the environmentally aware:

Does a 5-cent deposit on a soft-drink can help the environment? Mandatory deposits encourage recycling and reduce litter, but these programs typically spend $500 for every ton of cans and bottles collected, which makes curbside recycling look like a bargain. States without mandatory deposits—like Texas and Washington—have proven that the most efficient way to reduce litter is to hire clean-up crews, which pick up a lot more than just bottles and cans. Recycling takes money that could be used for other clean-up efforts: when New York's Sanitation Department started its recycling program, it cut back on street cleaning.

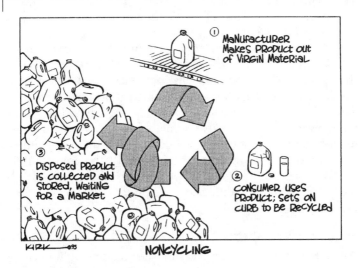

Reprinted by permission of Kirk Anderson.

Are reusable cups and plates better than disposables? A ceramic mug may seem a more virtuous choice than a cup made of polystyrene, the foam banned by ecologically conscious local governments. But it takes much more energy to manufacture the mug, and then each washing consumes more energy (not to mention water). According to calculations by Martin Hocking, a chemist at the University of Victoria in British Columbia, you would have to use the mug 1,000 times before its energy-consumption-per-use is equal to the cup. (If the mug breaks after your 900th coffee, you would have been better off using 900 polystyrene cups.) A more immediate environmental impact has been demonstrated by studies in restaurants: the average number of bacterial organisms on reusable cups, plates and flatware is 200 times greater than on disposable ones.

RECYCLING ACTUALLY CREATES POLLUTION

Should you recycle today's newspaper? Saving a tree is a mixed blessing. When there's less demand for virgin wood pulp, timber companies are likely to sell some of their tree farms—maybe to condominium developers. Less virgin pulp means less pollution at paper mills in timber country, but recycling operations create pollution in areas where more people are affected: fumes and noise from collection trucks, solid waste and sludge from the mills that remove ink and turn the paper into pulp. Recycling newsprint actually creates more water pollution than making new paper: for each ton of recycled newsprint that's produced, an extra 5,000 gallons of waste water are discharged.

Cost-benefit analyses for individual products become so confusing that even ardent environmentalists give up. After years of studies and debates about the environmental merits of cloth versus disposable diapers, some environmental organizations finally decided they couldn't decide; parents were advised to choose whichever they wanted. This sensible advice ought to be extended to other products. It would not only make life simpler for everyone, but would probably benefit the environment. When consumers follow their preferences, they are guided by the simplest, and often the best, measure of a product's environmental impact: its price.

Environmental Values

Polystyrene cups are cheap because they require so little energy and material to manufacture—without reading a chemist's analysis, you could deduce from the cup's low price that it's an efficient use of natural resources. Similarly, the prices paid for scrap materials are a measure of their environmental value as recyclables. Scrap aluminum fetches a high price because recycling it consumes so much less energy than manufacturing new aluminum. The low price paid for scrap tinted glass tells you that you won't be conserving valuable resources by recycling it. While price is hardly a perfect measure of environmental impact, especially in countries where manufacturers are free to pollute, an American product's price usually reflects the cost of complying with strict environmental regulations. It's generally a more reliable guide than intuitive moral judgments or abstract theories about what's good for the planet.

A theorist could logically argue that you have an obligation to recycle not just the paper in this magazine but also the staples. As a nonrenewable resource, isn't the steel theoretically even more precious than the paper? Shouldn't you take each staple to a scrap-metal dealer or, better yet, reuse it in your own stapler? But if you look at the low price of new staples—and the fact that scrap dealers aren't scurrying to buy used staples—you can see that it's a waste of time to worry about posterity running out of staples. Recycling devotees have too often ignored such signals, preferring programs based on rules instead of prices, and they've hurt their own cause.

"The sludge that results from municipal wastewater treatment processes contains organic matter and nutrients that . . . can improve the physical properties and agricultural productivity of soils."

RECYCLING SEWAGE SLUDGE INTO COMPOST IS SAFE AND EFFECTIVE

National Research Council

The following viewpoint is an excerpt of a report from the National Research Council. The council contends that treated municipal wastewater effluent, also known as sewage sludge and reclaimed water, can be safely used as compost on farmland and to irrigate crops. It maintains that current technology and regulations ensure that the risk of chemical contamination of food grown using reclaimed water and sewage sludge is negligible.

As you read, consider the following questions:

1. According to the council, what are some of the nonagricultural uses of reclaimed water?
2. In the author's opinion, why is the risk of human consumption of toxic chemicals from crops grown on sludge-treated land negligible?

Excerpted from the Executive Summary of the 1996 National Research Council report "Use of Reclaimed Water and Sludge in Food Crop Production." Reprinted by permission of the National Academy Press.

The use of treated municipal wastewater effluent for irrigated agriculture offers an opportunity to conserve water resources. Water reclamation can also provide an alternative to disposal in areas where surface waters have a limited capacity to assimilate the contaminants, such as the nitrogen and phosphorus, that remain in most treated wastewater effluent discharges. The sludge that results from municipal wastewater treatment processes contains organic matter and nutrients that, when properly treated and applied to farmland, can improve the physical properties and agricultural productivity of soils, and its agricultural use provides an alternative to disposal options, such as incineration, or landfilling.

MUNICIPAL WASTEWATER

Land application of municipal wastewater and sludge has been practiced for its beneficial effects and for disposal purposes since the advent of modern wastewater management about 150 years ago. Not surprisingly, public response to the practice has been mixed. Raw municipal wastewater contains human pathogens and toxic chemicals. With continuing advances in wastewater treatment technology and increasingly stringent wastewater discharge requirements, most treated wastewater effluents produced by public treatment authorities in the United States are now of consistent, high quality. When treated to acceptable levels or by appropriate processes to meet state reuse requirements, the effluent is referred to as "reclaimed water." Sewage sludge can also be treated to levels that allow it to be reused. With the increased interest in reclaimed water and the promotion of agricultural use for treated sludge, there has been increased public scrutiny of the potential health and environmental consequences of these reuse practices. Farmers and the food industry have expressed their concerns that such practices—especially the agricultural use of sludge—may affect the safety of food products and the sustainability of agricultural land, and may carry potential economic and liability risks.

Reclaimed water in the United States contributes a very small amount (probably much less than one percent) of water to agricultural irrigation, mainly because the extent of the practice is limited both by regional demands and the proximity of suitable agricultural land to many municipal wastewater treatment plants. Most reclaimed water goes towards various nonpotable urban uses such as irrigating public landscapes (parks, highway medians, lawns, etc.), air-conditioning and cooling, industrial processing, toilet-flushing, vehicle-washing, and construction.

Irrigation of residential lawns and/or gardens with reclaimed water is becoming increasingly popular where dual plumbing systems to facilitate water reuse have been installed; however, this report concentrates on agricultural uses of reclaimed water, and not residential use.

SEWAGE SLUDGE

Sewage sludge (or simply, "sludge") is an inevitable end product of modern wastewater treatment. Many of the organic solids, toxic organic chemicals, and inorganic chemicals (trace elements) are removed from the treated wastewater and concentrated in the sludge. An estimated 5.3 million metric tons per year dry weight of sludge are currently produced in the United States from publicly owned treatment plants. This amount will surely increase as a larger population is served by sewers and as higher levels of wastewater treatment are introduced.

Sludge disposal has always represented a substantial portion of the cost of wastewater management. Over the past 20 years, restrictions have been placed on certain sludge disposal practices (e.g., ocean dumping and landfill disposal), causing public wastewater treatment utilities to view the agricultural use of sludge as an increasingly cost-effective alternative. Currently, 36 percent of sludge is applied to the land for several beneficial purposes including agriculture, turfgrass production, and reclamation of surface mining areas; 38 percent is landfilled; 16 percent is incinerated; and the remainder is surface disposed by other means.

The Midwest has a long history of using treated sludge on cropland. Much of the cropland that receives sludge is used to grow hay, corn, and small grains for cattle feed, and public acceptance generally has been favorable. In Madison, Wisconsin, for example, the demand for sludge as a soil amendment exceeds the local supply. With ocean disposal of sludge no longer allowed, New York City and Boston—among other coastal cities—ship much of their sludge to other parts of the country. A portion of the sludge produced in the Los Angeles Basin is transported to a large farm near Yuma, Arizona.

AGRICULTURAL APPLICATIONS

If all the municipal sludge produced in the United States were to be agriculturally applied at agronomic rates, it would only be able to satisfy the nitrogen needs of about 1.6 percent of the nation's 1,250 million hectares (309 million acres) of cropland. About one quarter of this cropland is used to grow food for human consumption, of which 2 percent grows produce crops that can be consumed fresh. Thus, in a national context, the amount of food crops produced on fields receiving sludge would remain very small. Nevertheless, the local availability of agricultural land, combined with other regional and local concerns, is an important factor in sludge management decisions. While many western and midwestern states have ample agricultural land relative to the amount of sludge produced, land is less available in other regions. In New Jersey, for example, over half the state's cropland would be needed to receive sludge application to avoid other forms of disposal or out-of-state disposal. Rhode Island would need essentially all of its cropland to satisfy its sludge disposal needs through in-state agricultural use. The level of public acceptance for agricultural use of sludge varies considerably. Nuisance (e.g. odors and traffic), environmental, and safety issues are legitimate concerns that must be addressed by regulatory policy and management programs.

In February 1993, the U.S. Environmental Protection Agency

CORRECTING THE CASE FOR CAUTION

In early 1997, the Cornell Waste Management Institute (WMI) published a 44-page document titled *The Case for Caution: Recommendations for Land Application of Sewage Rules and an Appraisal of the US EPA's Part 503 Sludge Rules*. WMI suggests that land application of sewage sludge, or biosolids, should be more stringently regulated than demanded by the U.S. Environmental Protection Agency (EPA) in its Part 503 regulations (40 *Code of Federal Regulations*, Part 503). . . .

Reviewers of *The Case for Caution* have suggested that WMI is untrue to the scientific facts of the Part 503 risk assessment. . . .

Part 503 regulates nine metals sometimes found in biosolids. Other substances have been found in biosolids, but only so rarely and in such low concentrations that EPA judged these infrequent occurrences not to be a threat. Part 503 does not require monitoring of organics due to their minor presence in some biosolids. The potential for significant concentrations of organics and metals in biosolids has been reduced significantly by the implementation of pretreatment programs at the federal, state, and local levels. A recent survey by Vermont of virtually all municipal wastewater treatment facilities in the state found less than 3 percent organics in biosolids, with typical concentration rates below one part per million. Surveys by EPA and the Association of Metropolitan Sewerage Agencies have shown radioactivity concentration in biosolids to approximate environmental background levels.

Water Environment Federation, "Correcting *The Case for Caution*," May 1998.

(EPA) promulgated *Standards for the Use or Disposal of Sewage Sludge* (Code of Federal Regulations Title 40, Parts 257, 403, and 503, and hereafter referred to as the "Part 503 Sludge Rule"). This rule builds on a number of federal and state regulations that aim to reduce pollutants entering the municipal waste stream through source controls and industrial pretreatment programs that have reduced the levels of contaminants in the sludge as well as in the final effluent. The Part 503 Sludge Rule defines acceptable management practices and provides specific numerical limits for selected chemical pollutants and pathogens applicable to land application of sewage sludge. In this context, sewage sludge—traditionally regarded by many groups as an urban waste requiring careful disposal—is now viewed by the wastewater treatment industry, the regulatory agencies, and participating farmers as a beneficial soil amendment. . . .

SAFETY STANDARDS

Irrigation of food crops with treated municipal wastewater has been effectively and safely practiced in the United States on a limited scale. The public has generally accepted the concept of wastewater irrigation as part of larger and more comprehensive water conservation programs to reclaim wastewater for a variety of nonpotable uses. Where reclaimed water has been used for food crop production, the state standards for wastewater treatment and reuse, along with site restrictions and generally good system reliability, have insured that food crops thus produced do not present a greater risk to the consumer than do crops irrigated from conventional sources.

The beneficial reuse of municipal sludge has been less widely accepted. Federal regulations are designed to assure that sludge application for the production of food crops does not pose a significant risk from the consumption of foods thus produced. However, the parties affected by these reuse programs—local communities, crop growers, food processors, and the consumer—remain concerned about the potential for exposure to contaminants, nuisance problems, liability, and adequacy of program management and oversight. Sludge management programs based on agricultural sludge use can involve many potentially responsible parties, and can cross agency, state, and federal jurisdictional boundaries. Therefore, municipalities, public utilities, crop growers, and food processors must be able to provide well-managed and reliable programs that address, and are open to, community, business, health, agronomic, and environmental concerns. . . .

RECLAIMED WATER

States that regulate the use of reclaimed water for crop irrigation have focused on its microbiological quality and have not typically set human health criteria for harmful inorganic (trace elements) and organic chemicals in the reclaimed water. Instead, reliance is placed on the wastewater treatment processes to reduce these constituents to acceptable levels in reclaimed water.

Potentially harmful trace elements, such as arsenic, cadmium, cobalt, copper, lead, mercury, molybdenum, nickel, selenium, and zinc are found in treated municipal wastewater effluents. In 1973, the National Academy of Sciences issued a report on water quality criteria that recommended limits on the concentration of trace elements in irrigation water with regard to their effects on crop production. These agricultural irrigation guidelines have been generally accepted by EPA and others. Reclaimed water that has received a minimum of secondary treatment normally falls within these guidelines. While wastewater treatment processes typically used in the United States are not usually intended to specifically remove trace elements from the waste stream, most of the trace elements are only sparingly soluble and tend to become concentrated in the residual solids or sludge fraction. Chemical production and use bans, industrial pre-treatment programs, and municipal wastewater treatment have been effective in reducing the levels of toxic constituents in wastewater effluents to acceptable levels.

Wastewater treatment processes also remove many toxic organic chemicals in the wastewater stream through volatilization and degradation. Those that remain in the final effluent may volatilize or decompose when reclaimed water is added to soil. Consequently, only negligible quantities of toxic organic chemicals from municipal wastewater systems—those relatively resistant to decomposition—will persist in soils for an extended period. In general, toxic organic chemicals, especially those that persist in the soil, are not taken up by plants when the water application rates are commensurate with crop needs. Therefore, the immediate or long-term threat from organic chemicals to humans consuming food crops irrigated with reclaimed water is negligible.

TREATED MUNICIPAL SLUDGE

Potentially harmful chemicals (largely, trace elements and persistent organics) become concentrated in the sludge during the wastewater treatment process. Following repeated land applications, trace elements, except for boron, will accumulate in the soil to, or slightly below, the depth of sludge incorporation. The persistent organic chemicals degrade over time in soils. Degradation rates are dependent on the chemical in question and on soil properties.

The Part 503 Sludge Rule for the agricultural use of sludge sets criteria for concentrations of 10 trace elements in sludge; arsenic, cadmium, chromium, copper, lead, mercury, molybdenum, nickel, selenium, and zinc. The rule is based on a risk-assessment approach that considered the effects of these trace elements and organic chemicals of concern on crop production, human and animal health, and environmental quality. Except for cadmium, these trace elements are not ordinarily taken up by crop plants in amounts harmful to human consumers. EPA regulations for cadmium in sludge are sufficiently stringent to prevent its accumulation in plants at levels that are harmful to consumers.

In deriving pollutant loading rates for land application of sewage sludge, EPA considered 14 transport pathways and, in all cases, selected the most stringent value as the limit for each pollutant. For the 10 regulated inorganic pollutants, the most stringent loading rates were derived from pathways that involved a child directly ingesting sludge or from pathways involving effects on crops. This resulted in significantly lower pollutant limits than would have been the case had they been set by human food-chain pathways involving human consumption of food crops, meat or dairy products. Therefore, when sludges are applied to land according to the Part 503 Sludge Rule, there is a built-in safety factor that protects against human exposure to chemical contaminants via human food-chain pathways.

Available evidence indicates that most trace organic chemicals present in sludge are either not taken up or are taken up in very low amounts by crops after sludge is applied to land. The wastewater treatment process removes most of these organic chemicals, and further reduction occurs when sludge is processed and after it is added to soil. Consequently, only negligible quantities of toxic organic chemicals from municipal wastewater systems will persist in soils for an extended period. Recent studies suggest that plant tissues may absorb volatile toxic organic chemicals from the vapor phase of volatile compounds; however, the aeration that occurs during treatment of wastewater and during many sludge treatment processes removes most of the volatile organic chemicals at the treatment plant. . . .

A NEGLIGIBLE RISK

In summary, society produces large volumes of treated municipal wastewater and sewage sludge that must be either disposed of or reused. While no disposal or reuse option can guarantee complete safety, the use of these materials in the production of crops for human consumption, when practiced in accordance with existing federal guidelines and regulations, presents negligible risk to the consumer, to crop production, and to the environment. Current technology to remove pollutants from wastewater, coupled with existing regulations and guidelines governing the use of reclaimed wastewater and sludge in crop production, are adequate to protect human health and the environment. Established numerical limits on concentration levels of pollutants added to cropland by sludge are adequate to assure the safety of crops produced for human consumption. In addition to health and environmental concerns, institutional barriers such as public confidence in the adequacy of the regulatory system and concerns over liability, property values, and nuisance factors will play a major role in the acceptance of treated municipal wastewater and sewage sludge for use in the production of food crops. In the end, these implementation issues, rather than scientific information on the health and safety risks from food consumption, may be the critical factors in determining whether reclaimed wastewater and sludge are beneficially reused on cropland.

VIEWPOINT

> "Sludge can be dangerous on a
> number of counts. The most
> immediate danger are the pathogenic
> bacteria present in human waste,
> like the dreaded E. coli or
> salmonella."

RECYCLING SEWAGE SLUDGE INTO COMPOST IS A HEALTH HAZARD

Alex Todorovic

In the following viewpoint, Alex Todorovic argues that using sewage sludge as fertilizer is hazardous to human health and the future productivity of the land on which it is applied. Sewage sludge contains deadly chemicals and pathogens and therefore should not be applied to land where food crops are grown. Todorovic, the former news editor of the *Free Times*, South Carolina's largest weekly newspaper, writes frequently on environmental issues.

As you read, consider the following questions:

1. How much sewage sludge is produced in the United States each year, according to the author?
2. What is the difference among the three different classes of sludge, according to Todorovic?
3. How was sewage sludge disposed of before the current practice of applying it to land or dumping it in landfills?

Reprinted from Alex Todorovic, "Sewage Sludge: Toxic Brew or Miracle Fertilizer?" *South Carolina Free Times*, April 8–14, 1998, by permission of the author.

Before I began looking into sludge, I never much thought about where the swirling whirlpool at the bottom of my toilet led. Most people take plumbing for granted. Let's face it, once it's gone, who cares? And that's as it should be, but make an exception this once.

Let's take a ride to the holding tank at the Richland County, South Carolina, water treatment plant where bacteria are busy at work, breaking down our waste. Nine million gallons of the stuff passes through here each day where it eventually becomes sludge, a goopy mud-like substance.

Richland County turns its sludge into a soil conditioner by heat treating it to kill bacteria and then mixing it with sawdust. The end product is then sold to the public at the East Richland County Public Service District. What isn't converted into soil conditioner is placed in the landfill.

The use of sludge as fertilizer is coming under increased scrutiny by activists and a number of scientists because of its byproducts. Sludge is nutrient-rich and would be perfect for re-plenishing soil if it weren't for the fact that treatment plants also receive industrial toxic wastes, dirt off city streets, small doses of radiation from medical waste, pesticides, herbicides, and other leftovers from a poison-happy society which uses some 70,000 chemicals.

The resulting brew can be a toxic mix of nutrients, pathogens and poisons.

IT HAS TO GO SOMEWHERE

The United States produces 11.6 billion pounds of sewage sludge (dry weight) a year. All of this sewage sludge has to go somewhere. The options include incineration (as is practiced by the City of Columbia), landfill disposal, land application, and conversion into commercial fertilizer. Each one of these meth-ods has its shortcomings.

The trouble with incinerators is that they release high levels of pollutants into the atmosphere. Landfill disposal is safe, but expensive. Sludge is treated as toxic waste when dumped in landfills and there are strict (and expensive) rules governing its safe disposal.

The method gaining in popularity is to use sludge as a fertil-izer. In this scenario, private companies remove heat-treated sludge from a plant and find farmers who are willing to have it sprayed on their fields. "Free fertilizer!" they are told.

The least common method, conversion into a commercial soil conditioner, has its strong points, but the sludge must meet

"Exceptional Quality" standards which is a time-consuming process that involves killing large numbers of dangerous bacteria. Richland county sold 590.4 tons of this so called "Exceptional Quality" sludge-made soil conditioner in 1997 alone. Most of this went to landscape companies.

Is It Safe?

Sludge can be dangerous on a number of counts. The most immediate danger are the pathogenic bacteria present in human waste, like the dreaded E. coli or salmonella.

There are three levels of treatment for sewage sludge which produce three grades of sludge, "Exceptional Quality," "Class A" and "Class B" with Class B being the least treated kind. There are different rules for using Class A and Class B sludge as fertilizers. Fields that have been sprayed with the poorest quality sludge are so contaminated with pathogens that the regulations say, "Public access to land shall be restricted for one year after application of sewage sludge."

Although using sludge as fertilizer isn't a new practice, it has exploded in recent times. It has been a mixture in a number of commercial fertilizers since 1927, but is now being applied directly on crop lands.

Today, activists and a number of scientists are sounding the alarm about this practice. Ironically, spraying crops with sewage sludge began when an activist group, Clean Ocean Action, launched a successful campaign to protect the oceans from toxic dumping. Large cities used to dump their sewage straight into the ocean until Clean Ocean Action challenged this practice. The EPA and municipalities wanted to continue ocean dumping of sludge, but the activists prevailed. By the early 1990s, cities had to find new disposal methods for their sewage sludge.

The Sewage PR Problem

In the late 80s, municipal governments and the sewage treatment industry could see the changing tide and they came up with a new plan; to spread sludge on land, preferably close to its origination point to lower transportation costs.

Since the public had a generally bad impression of toxic sewage sludge, its use as a fertilizer would require a major public relations makeover—from toxic pollutant to a beneficial substance. The waste-treatment industry hired public relations firm Powell Tate to clean up sludge's image, and a contest was held to rename the noxious matter. Entries included "sca-doo," "hu-manure," "hu-doo," and "bioslurp." But it was "biosolids" which took first prize and "biosolids" it still is. Most people still say sewage sludge.

THE SCIENTIFIC BLACK HOLE

The EPA tried to assess the risk of using sludge as a fertilizer before the new regulations were drawn up in 1993. But the EPA's methodology to assess risk has come under fire. In 1988, EPA sampled sludge from 180 sewage treatment plants, but they only looked for 409 chemicals—the United States uses roughly 70,000 chemicals. Critics say that "detection limits" for many organic chemicals were set so high that few were detected even though many were present.

Reprinted by permission of Kirk Anderson.

Of the original 409 chemicals examined, the EPA narrowed the list to 28, which were labeled "of concern." From that list of 28, EPA picked 10 metals that they would regulate: arsenic, cadmium, chromium, copper, lead, mercury, molybdenum, nickel, selenium, and zinc. The EPA asked how much of each of the 10 pollutants a "highly exposed individual" would be exposed to in various scenarios. It did not, however, examine the overall impact of the metals on crops. The EPA called its study a "comprehensive" risk assessment.

Complicating matters is the fact that much of the science on sewage sludge has been "friendly science," funded by the waste industry and designed to break down public resistance to the use of sewage sludge on farm land. There is a black hole of information when it comes to the long-term use of sewage sludge as a fertilizer.

European countries use sewage sludge as fertilizer, but they have taken a more cautious approach, setting much stricter limits on how much sludge can be used.

A recent study by Cornell University's Waste Management Institute entitled, "A Case for Caution," also points to the fact that we've rushed into something without knowing the long-term consequences. "Sewage sludge has nice nutrients," explained Ellen Z. Harrison, director of Cornell University's Waste Management Institute. "The thing we sometimes forget is that it also is industrial waste."

Harrison and her Cornell colleagues say that the EPA standards are not adequate to protect public health or the long-term health of soil. They contend that the cumulative loading limits are roughly ten times too high.

Harrison points out that other countries have done similar risk-based research and come up with much stricter standards. "Their system was based on eco-toxicity and not just human health."

The Cornell study generated a storm of controversy around an already politicized issue. "The study made a lot of people upset, and a lot of other people pleased," said Harrison. "Our objective was not to do either of those things. Our mission was to show that there is a scientific rationale for taking a more cautious approach. Our position has been characterized as anti-sludge. That's not what we are saying. You can use it, but we should take a more cautious approach."

University of California-Davis professor Bill Liebhardt also says we don't know enough yet. "How much lead and cadmium is going to get in a particular crop—all you can say is it depends on a lot of factors," he said. "Some crops may not take up hardly any of it, and other crops may take up quite a bit."

Even more alarming than sewage sludge is the legal practice of spreading industrial waste on farm land in the guise of "fertilizer." In 1997, the *Seattle Times* documented this disposal method in a series of articles entitled "Fear in the Fields." There have already been documented instances of cow and crop poisonings resulting from this disposal method.

AN IMAGE PROBLEM

To this day, the waste industry struggles with its public image. Charlotte-based Bio-Nomic Services has a contract to land-apply

Class B sludge on 20,000 acres of South Carolina land. North and South Carolina have an agreement whereby sewage sludge can be carried over state lines and land-applied.

Bio-Nomics has 19 offices and has been land-applying sludge since 1991. The company uses Class B sludge because it is the most economically favorable.

Mike Pontabello and David Motil, both of Bio-Nomics, acknowledged that their industry is locked in a PR fight. "In Virginia, you have a huge public perception problem. They even have a cable show about it," said Motil.

They are also thankful that, so far, the PR battle has stayed out of the Carolinas. "We could use some good press," they both said.

PERIODICAL BIBLIOGRAPHY

The following articles have been selected to supplement the diverse views presented in this chapter. Addresses are provided for periodicals not indexed in the *Readers' Guide to Periodical Literature*, the *Alternative Press Index*, the *Social Sciences Index*, or the *Index to Legal Periodicals and Books*.

Tom Arrandale	"The Big Burnout," *Governing*, August 1998.
David Bacon	"Recycling—Not Always Green to Its Neighbors," *Neighborhood Works*, May/June 1998.
Jeff Bailey	"The Recycling Myth," *Reader's Digest*, July 1995.
Leora Broydo	"The Vinyl Analysis," *Mother Jones*, March/April 1998.
Barry Commoner	"Recycle More and Spend Less," *New York Times*, July 6, 1996.
Anne Marie Cusac	"Nuclear Spoons," *Progressive*, October 1998.
David Fischer	"Turning Trash into Cash," *U.S. News & World Report*, July 17, 1995.
Leslie A. Goodwin	"It's Not Easy Bein' Green in a World That Likes to Waste," *Christian Science Monitor*, May 21, 1996.
Cathy Madison	"Don't Buy These Myths," *Utne Reader*, November/December 1998.
Arthur H. Purcell	"Trash Troubles," *World & I*, November 1998. Available from 3400 New York Ave. NE, Washington, DC 20078-0760.
Gunjan Sinha	"Electric Landfills," *Popular Science*, August 1998.
Roger Starr	"Recycling: Myths and Realities," *Public Interest*, Spring 1995.
Clark Wiseman	"Recycling Revisited," *PERC Reports*, August 1997. Available from 502 S. 19th Ave., Suite 211, Bozeman, MT 59718.
Joyce Yeung	"How to Recycle a House," *Mother Earth News*, February/March 1998.

CHAPTER 4

HOW CAN AIR POLLUTION BE REDUCED?

CHAPTER PREFACE

Although smog and ozone levels have declined dramatically since the early 1990s, Americans are still breathing remarkably polluted air. Incredibly, this air is not in congested cities, but inside homes and office buildings. The Environmental Protection Agency (EPA) reported in December 1998 that airborne pollutants were two to five times higher indoors than out. According to the EPA, the air inside is less healthy than the air outside because emissions from automobiles and industrial smokestacks have declined and because buildings are now built so tightly and so well-insulated that indoor air-borne pollutants are unable to escape.

The indoor pollutants recorded by the EPA are found in most American homes. They include polycyclic aromatic hydrocarbons from kerosene heaters and fireplaces, and nitrogen oxides from gas stoves. Also listed are secondhand smoke from cigarettes and cigars; mold and mildew spores; dust mites; chemical fumes from paint, plywood, and compressed wood; styrene and 4-PC from carpet backing and adhesives; paradichlorobenzene, an animal carcinogen in mothballs and deodorizers; and the dry-cleaning solvent perchloroethylene. Other fumes are released into the air by aerosol sprays, cleaners and solvents, and paint strippers, to name a few. Many of these indoor pollutants are known to irritate the lungs, bring on asthma attacks, and increase the risk of chronic diseases and cancer.

Environmentalists suggest that the problem of indoor air pollution can be greatly ameliorated by opening doors and windows to let in fresh air. In the winter, when opening doors is impractical, they recommend using trigger sprays instead of aerosols; using bleach and dehumidifiers to kill mold and bacteria; airing out dry-cleaned clothes and new carpeting before bringing them in the house; sealing raw wood edges to prevent the release of noxious fumes; and ventilating stoves and heaters outside.

Reducing smog and particulate matter in the air outside is not as easily accomplished as improving the air quality indoors. The authors in the following chapter examine whether stricter air quality regulations, pollution credits, or electric cars will reduce air pollution.

| "The overwhelming body of evidence told us that air-pollution standards are not adequate to protect the public's health and that the current standards leave many at risk."

STRICTER REGULATIONS WILL REDUCE AIR POLLUTION

Carol M. Browner

Carol M. Browner is the administrator of the Environmental Protection Agency (EPA) and the former head of Florida's Department of Environmental Regulation. In the following viewpoint, Browner argues that stricter air quality standards are needed to protect the nation's health. She maintains that compelling scientific evidence shows that the current air pollution standards leave many people, especially children, at risk for lung ailments and even death. The health and safety of the nation is the most important factor in determining the new standards, not the projected costs of revising the standards, Browner asserts.

As you read, consider the following questions:

1. What steps must the EPA go through to ensure that the current and revised air quality standards protect the public health, according to Browner?
2. What evidence does Browner present to support her contention that air pollution contributes to ill health and death?
3. In the author's opinion, when is it appropriate to consider the costs of new air pollution standards?

Since the 1970s, America has made great progress in protecting the public's health and the air, the water and the land that we all share. As one example, the air in most major metropolitan areas is cleaner than it once was. This is due to the fact that together, as a nation, we have insisted on strong public-health protections—and because businesses, communities and government agencies have worked in commonsense and cost-effective ways to meet the health standards we have set.

STRENGTHENING THE STANDARDS

In November 1996, the Environmental Protection Agency, or EPA, proposed to strengthen the national ambient-air quality standards for particulate matter and ground-level ozone (better known as soot and smog). We know from the best, current science that strengthening the ozone standard would protect tens of millions of Americans—including millions of children—from the adverse health effects of smog. And science tells us that the new standards we proposed for particulate matter would result in many thousands fewer cases of premature death, aggravated asthma and acute respiratory distress in children each year.

To be sure, these proposed air standards have generated much controversy. . . .

Some critics, I'm afraid, have gone beyond the pale—branding the EPA as "extremist" and using scare tactics about intolerable lifestyle changes that these proposed standards would force on the American people. What they haven't talked about is the public's health and how to protect it.

And that brings up an interesting question—namely, in matters of public health, how much weight should be given to vested interests? How much say should regulated industry have in the setting of air-quality standards?

THE CLEAN AIR ACT

Fortunately, the Clean Air Act provides some answers. Born under President Richard Nixon, amended and strengthened under Presidents Jimmy Carter and George Bush, the Clean Air Act is the embodiment of an ongoing bipartisan desire to protect all Americans from the harmful effects of breathing polluted air.

From the beginning, the act has contemplated the march of technology and science. It has recognized that science always will come up with better ways to understand the health effects of the air we breathe—and that the standards of the seventies may not be right for the 21st century. And it anticipated that some with vested interests in the status quo might not agree with the conclusions of the scientific and public-health communities.

Therefore, Congress set forth a process to ensure that the standards would be set and, if necessary, revised in a manner that puts the public health first and ensures that Americans are protected with an adequate margin of safety.

First, the Clean Air Act directs the EPA to review the public-health standards for the six major air pollutants at least every five years to ensure that they reflect the best current science. It also lays out a specific procedure to obtain such science and, if needed, revise the standards. This is to ensure that we never get to the point where the government tells Americans that their air is healthy to breathe when, in fact, it is not.

Jobs and Profits

Predictably, polluters have claimed that the proposed revision of air quality standards will trigger job losses. But industry costs don't automatically translate into such losses. In fact, pollution-control requirements have caused a small net increase in jobs because they tend to force polluters to hire people to control pollution. Furthermore, we shouldn't equate a polluter's economic interests with those of society as a whole. Polluters often oppose pollution control precisely because they prefer high profits to increased hiring.

David M. Driesen, *Christian Science Monitor*, March 3, 1997.

Next, the process requires that EPA's standard-setting work and the underlying health studies—250 of them in the case of ozone and particulate matter—be independently reviewed by a panel of scientists and technical experts from academia, research institutes, public-health organizations and industry. The independent panels for ozone and particulate matter conducted 11 meetings, all open to the public—a total of 124 hours of public discussion of the scientific data, research and the studies of the health effects of smog and soot.

EPA has held further public meetings at which hundreds of representatives from industry, state and local governments, organizations—as well as members of the public—have offered their views. We are in the process of analyzing and considering the submitted comments—a process we take very seriously and one that will weigh heavily on the final outcome. If, in the end, that process leads the EPA to set new public-health standards, Congress then has its say and may vote up or down on them.

Congress, of course, has been watching this process closely.

And Congress does have a review process for these kinds of regulations. It is a good process—as long as it is carried out responsibly, looks at all the evidence and judges it in the best interests of all Americans. The EPA and the independent scientific panels did look at the evidence—all of it published, peer-reviewed and fully debated. The overwhelming body of evidence told us that air-pollution standards are not adequate to protect the public's health and that the current standards leave many at risk. The EPA therefore has proposed to tighten the standards to ensure we are being truthful with the American people about the quality of the air they are breathing and what it is doing to them.

COMPELLING EVIDENCE

The science on which we based our proposal is compelling. It shows that smog aggravates asthma and other respiratory ailments and causes temporary decreased lung function in up to 20 percent of healthy people. One study showed large increases in hospital admissions when smog levels were up. Even on days when smog levels were at or below the current U.S. standard, nearly one in three hospital admissions for respiratory problems was linked to smog. On high-pollution days, smog was associated with about half of all respiratory admissions.

Another study found that nonsmoking men and women living in areas with relatively high levels of smog had more than half the lung damage of a pack-a-day smoker. Still another study indicates that healthy young adults engaging in routine construction work outdoors will experience significant breathing problems at ozone levels equal to the current standard. And a study of active children at an outdoor summer camp found consistent loss of lung function at or below the current ozone standard.

For particulate matter—soot—the studies consistently have found a correlation between this pollutant and early death. One analysis, published in 1993 in the *New England Journal of Medicine*, showed that exposure to fine particulates in the air increases the risk of early death by 26 percent and, in the most polluted cities, shortens lives by an average of one to two years. A study in Birmingham, Ala., showed that air with higher soot levels directly was associated with higher numbers of elderly people going to the hospital for respiratory and cardiovascular illnesses.

At a time when lung disease is the third leading cause of death in America and when asthma has become the most common chronic illness in children, is it not prudent that we heed what scientists and public-health officials are telling us about our air?

No less an authority than the American Academy of Pediatrics has recommended that pediatricians make parents aware of the daily variations in ozone and, when ozone levels are high, keep their kids indoors.

I do not believe this is the kind of future Americans are willing to accept. Americans want clean air. They want their children protected. They want the EPA to do its job—which is ensuring that the air they breathe is safe. They want government to be honest when the air is unhealthy. And they have every right to expect that industry will rise to the occasion, meet the challenge and once again reduce their pollution of the public's air, if that is required.

THE COST OF CLEAN AIR

Let me be clear, when it comes to implementing these standards, we consider the costs. But the law does not, and should not, allow us to consider costs at this critical public-health stage of the process. The Clean Air Act clearly requires levels of smog and soot to be based solely on health, risk, exposure and damage to the environment, as determined by the best available science—and not projected costs for reducing pollution.

Looking back, nearly every time we have begun the process to set or revise air standards, the costs of doing so have been grossly overstated—by both industry and the EPA. Dire predictions of economic chaos, always a part of the clean-air debate, have never come to pass. Why? Because industry ultimately rose to the challenge of finding cheaper, more innovative ways to meet the standards—and reduce its pollution.

However, when it comes to implementing clean-air standards, it is appropriate to weigh the costs, industry by industry, of reducing pollution—allowing us to find the most effective solution. If, after review and analysis of the public comment, these new standards are adopted, the EPA will work with all who are affected—state governments, local governments, community leaders, businesses large and small—to find cost-effective and commonsense strategies for reducing pollution and providing the public-health protections. This is part of the process, too. To further this effort, the EPA has created an advisory subcommittee—comprising nearly 60 representatives from state and local governments, industry, small businesses and environmental groups—to propose innovative approaches for implementing any new standards for soot and smog. In addition to the subcommittee, there are five working groups comprising about 100 more representatives of these same kinds of organizations and enterprises.

When I first announced the proposed revisions to the air standards in November 1996, I directed the EPA to further expand the membership of the advisory subcommittee to include more representation from small businesses and local governments. Also, in conjunction with the Small Business Administration and the Office of Management and Budget, the EPA has been holding meetings with representatives of small businesses and local governments to obtain their views on these proposed standards.

Working together, we can have cleaner air and not sacrifice our economic vitality. We've done it before. Since 1970, emissions of the six major air pollutants have dropped by 29 percent, while the population has grown by 28 percent and the gross domestic product has nearly doubled.

Economic growth and cleaner air: Now, that is a level of progress we all can be proud of. Experience tells us that the two go hand-in-hand. Let us respond as we have before. Let us listen to the science, and let us work together toward the goal of cleaner and healthier air.

> "The newly proposed standards would purportedly convey additional protection to only limited groups, many of whom can modify their risk-exposure through behavioral change."

STRICTER REGULATIONS MAY BE COSTLY AND INEFFECTIVE

Kenneth Green

The air quality standards for ozone and particulate matter do not need to be tightened, argues Kenneth Green in the following viewpoint. There is concern among scientists that the Environmental Protection Agency's recommendations are based on flawed studies. Furthermore, Green asserts, the expected costs of implementing the new standards will exceed any benefits. Green is the director of environmental studies at the Reason Public Policy Institute, a libertarian organization in Los Angeles, California.

As you read, consider the following questions:

1. According to Green, how many counties are expected to be nonattainment areas if the proposed EPA regulations are passed?
2. What are the estimated costs of compliance with the new EPA standards according to the EPA and according to Alicia Munnell, as cited by the author?
3. Why are the proposed EPA standards unfair to some groups, in Green's opinion?

Abridged from Kenneth Green, "Proposed Ozone and Particulate Matter Standards," Issue Brief, March 26, 1997. Reprinted by permission of the Reason Foundation.

On November 27, 1996, the Environmental Protection Agency (EPA) announced its intention to tighten two already challenging air quality standards; one which regulates exposures to ground-level ozone, and one which limits exposures to small particulate pollution known as particulate matter (PM). EPA's justification for performing the review at this time is straightforward: the agency was being sued by several environmental and public health advocacy groups calling for such a review. EPA's justification for the proposed standards that emerged from the review process is equally straightforward—according to their review of the scientific literature, EPA asserts that the existing standards for ozone and PM are not fully protective of the public's health. Meeting the proposed new standards will pose a stiff challenge in over 500 counties across the United States, and the proposed standards raise many questions warranting further exploration. . . .

PROPOSED CHANGES TO THE OZONE STANDARD

Ozone is a colorless, odorless gas produced by a variety of chemical reactions (at ground-level) involving nitrogen oxides and volatile organic compounds in the presence of sunlight. Ozone is a known respiratory irritant, implicated in causing decreased lung function, respiratory problems, acute lung inflammation, and impairment of lung defense mechanisms. Outdoor workers and active children are claimed to be particularly at risk.

EPA has proposed lowering the existing standard for ozone from 0.12 parts per million measured over one hour to 0.08 parts per million, measured over eight hours. At the time of the announcement, 106 counties (in 26 states) were in violation of the existing standard, and estimates indicate that the proposed new standard would increase the number of such "nonattainment areas" by 229 counties (adding counties in seven additional states), bringing the total to 335 counties in 33 states.

PARTICULATE MATTER

Particulates are a class of airborne pollutants ranging in size from less than one micron (one thousandth of a millimeter) in diameter up to 100 microns (a bit wider than the average human hair). Particulates are generated through a range of natural and man-made processes including combustion and physical-abrasion. Small particulate matter, under 10 microns in diameter, has been implicated in increased mortality for the elderly, as well as those members of the population with damaged respiratory systems. Small particulates have also been claimed as aggravating factors for existing respiratory and cardiovascular disease, resulting in more frequent and/or serious attacks of asthma in children.

The existing particulate matter standard treats all particulates smaller than ten microns as a single class, called PM10. Under the existing particulate standards, ambient air concentrations of PM10 are limited in 24-hour average concentration to 150 micrograms per cubic meter ($\mu g/m^3$) of air, and in average annual concentration to 50 $\mu g/m^3$ of air. Because recent studies have implicated particles under 2.5 microns in diameter (PM2.5) as being especially harmful, EPA proposes to single that class of particulates out for additional regulatory attention. The proposed new standards would retain the same concentrations for particulates ranging from 2.5 to 10 microns in diameter, but would add two new standards for particulates below 2.5 microns. The new PM2.5 standards would limit 24-hour ambient concentrations to 50 $\mu g/m^3$ and the annual average to 15 $\mu g/m^3$.

Under the existing PM10 standard, 41 counties in 19 states are designated nonattainment areas. The proposed new standards would drop the PM10 nonattainment counties to 11 (in 5 states), but would add 167 PM2.5 nonattainment counties in 37 states.

OUTSTANDING QUESTIONS

Many questions remain regarding the proposed new standards, ranging from questions about the scientific rigor of the review process, to economic and equity issues. Given the potential impact of the proposed rule, thoughtful examination of these questions seems warranted.

How good is the science? EPA's claims that the current standards are not protective rest on complex risk analyses that EPA has not made fully available to the public for independent review. That's troubling, since scientists, both inside and outside of the standard-review process, have expressed concerns that the science behind the risk assessments might be flawed, and may not, in fact, point clearly to a need for tighter standards. For ozone, a well-understood pollutant, the primary concerns involve determining what standard is considered protective. Uncertainty in this regard raises the specter of selecting a value in haste, and repenting at leisure through mid-course revisions. In the case of particulate matter, more data is missing than is known. Knowledge is severely limited regarding particulate matter measurement, composition, hazardous constituents, control feasibility, and so on.

How much will it really cost to comply? Estimated compliance costs of the new standards are high: EPA estimates the national cost of new compliance efforts triggered by the new standards at $6.5 billion to $8.5 billion annually. This estimate seems low to some analysts, considering that California's South Coast Air Basin is planning to spend $1.7 billion per year just to meet the existing standard, while proposed efforts in the Chicago area are estimated to cost $2.5 to $7 billion annually. Alicia Munnell, a member of the President's Council of Economic Advisers estimated the costs of attaining the new ozone standard alone at $60 billion annually. While EPA asserts that it is not allowed to consider cost as a factor in assessing risk (an assertion discussed further below), it's clear that cost matters. Few people would probably agree with "clean air at all costs," if it turned out that we would get the most trivial reductions in emissions at the cost of a sound economy. . . .

AMBIENT AIR QUALITY STANDARDS

Current and Recommended Ambient Air Quality Standards for PM10*

	Environmental Protection Agency Current	Proposed	National Resources Defense Council Recommendation	American Lung Association Recommendation
24-hour average	150	150	33	50
Annual average	50	50	17	30

Current and Recommended Ambient Air Quality Standards for PM2.5*

24-hour average	NS	50	20	18
Annual average	NS	15	10	10

*micrograms per cubic meter of air

NS = No Standard

Michael Fumento, *Reason*, August/September 1997.

EPA interprets its obligations under the Clean Air Act to require them to ignore this information, however, and perform analyses of health impacts without regard to the potential for risk-shifting as a side-effect of its proposed standards. Indeed, the Clean Air Act forbids consideration of economic or technologic feasibility, but, as far as we've been able to determine thus far, the statute is mute on the question of factoring in side-effects of a proposed rule using economic information as a surrogate measure of harm. It is EPA's interpretation of the rule that seems to govern the exclusion of such information from the risk assessment phase of the standard-review process. But this interpretation seems illogical; akin to a physician considering a prescription without regard for side-effects, or multiple drug interaction. This interpretation is also highly selective, given that EPA has long championed consideration of exactly such risk-shifting side-effects when evaluating the environmental impacts of activities such as automobile use. It seems reasonable to expect EPA to apply this sound thinking uniformly, rather than selectively. What are the potential side-effects of the proposed new standards? How much health benefit do they subtract from EPA's claimed health improvement? How does EPA account for the well-documented risk-increasing effects of poverty, unemployment, reduced funding for law enforcement, reduced educational opportunity, or poorer nutrition? Do EPA's benefit calculations account for our reduced ability to simultaneously address higher environmental risks, such as those purportedly posed by poor indoor air quality? When all possible side-effects of the proposed new standards are taken into account, will they provide any additional protection at all, or could they actually lead to increased risk from side-effects greater than the claimed harm?

FAIRNESS

Are the proposed standards fair to different social groups? Previous ozone and particulate standards were designed to protect all of a non-attainment area's population from air pollution concentrations that were harmful to virtually all who were exposed, and where minimizing exposure was nearly impossible at the high pollutant concentrations of the past. The air has gotten cleaner, however, and the newly proposed standards would purportedly convey additional protection to only limited groups, many of whom can modify their risk-exposure through behavioral change. And there are equity issues to consider as well. Low-income groups suffer more from economic downturns than higher-income groups—how does this affect their potential benefit from the proposed standards? Some communities have such high rates of deadly violence that diverting limited resources to reducing particulate concentrations is akin to straightening the deck chairs on the Titanic. Is a one-size-fits-all standard compatible with local risk-reduction priorities?

Certainly we must protect our children, who have limited ability to affect their exposure to air pollution risks—virtually nobody questions that need. But we can legitimately question whether national standards are either the best way, or the fairest way, to protect those who need protection.

Are the proposed standards fair to different geographic regions? Not all areas of the country will be impacted in the same way by the new ozone and particulate matter standards. The Southwest and Northeast, already struggling with the existing standard, will be particularly hard hit by the proposed new standards. This is significant because of the high costs of compliance projected for the proposed new standards—not only in terms of dollars, but in terms of extensive lifestyle impacts. Is it fair to impose standards upon the entire country that some regions simply cannot attain due to geographic, demographic, or other factors beyond their control?

Many regions that are out of attainment for the existing standard find themselves unable to comply despite extensive regulations and significant levels of spending. It is only logical to ask whether there is any sense in setting a standard that one can't meet. Before setting a new standard, a determination of feasibility should be investigated.

WHERE DO WE GO FROM HERE?

There are enough unanswered questions regarding the effectiveness, efficiency, and fairness of EPA's proposed new standards for ozone and particulate matter to justify further study. Rather than rush the proposed new standards into law, creating 500 separate battles over implementation, EPA should allow time for meaningful study of these issues with the goal of avoiding wasted effort. Enactment of the proposed new standards amidst such uncertainty is likely to generate conflict and distrust, weakening the legitimacy of our nation's environmental laws.

"Trading [pollution credits] allows companies to deal with pollution that otherwise would be too expensive to reduce."

POLLUTION CREDITS WILL REDUCE AIR POLLUTION

John J. Fialka

In 1995 the Environmental Protection Agency lowered the acceptable acid-rain emissions rate for airborne pollutants by public utility companies and at the same time awarded the utilities certificates that permitted them to exceed their emissions allowances. Utility companies that met or exceeded their emissions goals were encouraged to sell their pollution credits to companies that were unable to meet the lowered standards. In the following viewpoint, John J. Fialka reports that the program has been so successful in reducing the amount of acid-rain emissions that the program is being considered as a means of reducing carbon dioxide emissions worldwide to stop global warming. Fialka is a staff writer for the *Wall Street Journal*.

As you read, consider the following questions:

1. According to the Environmental Protection Agency, by how much has the sulfur dioxide emissions dropped due to the pollution credit program?
2. What was the average price for a pollution certificate in 1997, as cited by Fialka?
3. What countries are expected to buy their emissions allowances for carbon dioxide, according to the author?

As part of a sixth-grade science project, Rod Johnson's students in Glens Falls, N.Y., removed 330 tons of sulfur dioxide from the air.

Holding raffles, bake sales and auctions over a three-year period, they raised $25,000 to buy 330 certificates in the U.S. Environmental Protection Agency's acid-rain-emissions trading program—one of the nation's hottest new commodity markets. Each certificate allows the owner to emit one ton of the noxious gas.

Utilities trade the certificates—some selling them for profit, others buying them to comply with air-quality standards. But the students at Glens Falls Middle School are going to sit on theirs. The way the EPA program works, that means the nation's air will be that much cleaner.

Proposed by the Bush administration in 1990 as a novel, market-oriented solution to the problem of acid rain, the trading of what amounts to sulfur-dioxide-pollution permits has led to results that have exceeded expectations. Since its 1994 inception, the trading program, administered by the EPA, has contributed to a 30% drop in sulfur-dioxide emissions from major polluters, the agency says.

A Cheap Alternative

President Bill Clinton hopes to sell a version of the trading program to other nations meeting in Kyoto, Japan, in December 1997 to conclude a treaty that curbs global warming. Though some environmentalists don't like this approach, and government agencies are finding it hard to explain the complex system to local citizens, business-people are warming to it. And Mr. Johnson says his students like trading because, "We can identify a problem and then participate in a solution."

Trading appears to be a cheap way to help curb pollution. Industry had complained that removing sulfur dioxide from the air would cost as much as $1,500 a ton. But the price of about 7.2 million certificates being traded in 1997 reflects a cost of $90 a ton. After years of handwringing, acid rain is being reduced by a market that lets individual companies make their own antipollution strategies.

"This program turns companies into pollution minders," says Daniel J. Dudek, a senior economist for the Environmental Defense Fund in New York, who sold the idea to the Bush administration and has become a kind of Johnny Appleseed, spawning other trading plans. "When they [polluters] think about it, they say, "Hey! That's money going up the stacks."

Trading allows a company to hunt for the most cost-efficient ways to reduce emissions. In some programs it doesn't necessarily have to be their emissions. The Los Angeles area's South Coast Air Quality Management District, for example, established a program in which it gave four oil refineries credit in the form of certificates for removing the equivalent 2,000 tons of smog after they bought and junked 17,502 pre-1982 cars, which, the agency says, spewed an equivalent amount of pollution over the four-county district.

But the program has been extremely difficult to sell to residents of Wilmington, an unincorporated area near Los Angeles Harbor where homes and schools sit next door to several refineries. The eyes of Elidia Madrigal, a teacher's aide in the local elementary school, widen as a concerned neighbor explains the program to her in Spanish: "If they take a couple of bags of garbage away from the street, it allows them to throw a couple of bags in front of your house," the neighbor says.

"This is unfair! We're not going to tolerate this," Ms. Madrigal says, complaining that the fumes from a refinery half a block away sometimes make her and her students sick. A local activist group, Communities for a Better Environment, is suing the refineries, the state and the Air Quality District in Federal District Court in Los Angeles, charging that the trading program violates the civil rights of residents of the predominantly Mexican-American neighborhood. They want to stop the trading program and get the government to set tighter emissions standards.

But Paul Eisman, a senior vice president for Ultramar Diamond Shamrock Corp., which owns the nearby refinery, loves the program. By buying and junking 334 old cars, it was able to avoid installing more expensive vapor-control equipment.

"This is a microcosm of a huge issue," says Mr. Eisman, who adds that trading allows companies to deal with pollution that otherwise would be too expensive to reduce. "The emission that's least costly to reduce gets reduced first," he says. But he adds that he understands Ms. Madrigal's concern: "If you're living next door to a situation like this, you're never going to be totally happy with it."

Kevin Snape, director of the Clean Air Conservancy, a Cleveland-based environmental group that promotes trading solutions, says local "hot spots" will disappear as cheap certificates are bought up and companies discover it is less expensive to buy new antipollution equipment.

In Ohio, he notes, utilities have found that byproducts of trading in the federal acid-rain program include cheaper and more effective scrubbers and new low-sulfur fuel-mixing techniques. "All of this would have never happened under the old regulatory model that said you must install this specific technology," he says.

THE BASIC IDEA

The basic idea for the sulfur-dioxide program is relatively simple. The government set a cap that reduced the number of tons of the pollutant it would allow in the air, starting in 1995 and declining thereafter. Then it issued to 110 of the nation's dirtiest power plants tradable certificates that matched their share of the cap. Companies that cut emissions below their cap had extra certificates to sell. Companies that didn't had to buy them from the cleaner companies. Companies can save certificates from year to year, but federal clean-air standards still limit the amount of pollution that can be released.

For Milwaukee-based Wisconsin Electric Power Co., a unit of Wisconsin Energy Corp., the cap meant it had to cut sulfur-dioxide emissions from its five power plants by about 30,000 tons. Daniel L. Chartier, the company's emissions manager, figured he could remove 20,000 tons relatively cheaply by switching to low-sulfur coal. To get the remaining 10,000 tons, he learned, would be tough. He would have to buy two house-size, $130 million machines called "scrubbers."

But because other companies had reduced their emissions well below the cap, the market was flooded with cheap allowances. And since buyers can approach sellers in any part of the country, prices tend to even out. Mr. Chartier bought 10,000 of them, estimating he saved his company more than $100 million. In 1997, he formed the Emissions Marketing Association and 63 companies, mostly large utilities, have joined it. "We found growing enthusiasm as the knowledge of emissions trading increased," he says.

More markets are coming. Twelve northeastern states are adapting a version of the federal sulfur-dioxide trading program to reduce ozone levels, starting in 1999. Some of them have also joined a group of 37 states considering a larger smog-reduction plan. California has hatched a variety of such programs.

While some may find it hard to envision pollution rights as a commodity, trading is already a big business. The EPA expects sulfur-dioxide trading to reach $648 million in 1997. The utility industry estimates actual volume is probably double that—or equal to the $1.2 billion cash market for U.S. soft red winter wheat. "Trades have basically doubled every year," says Brian McLean, head of EPA's acid-rain division, which tracks trades.

The Environmental Defense Fund's Mr. Dudek says this is only the beginning. He is consulting with traders at British Petroleum Co., helping them determine how a world-wide market in carbon-dioxide-emission allowances would work.

THE GLOBE'S DIRTY DOZEN

Annual carbon-dioxide emissions

	Total Tons (millions)	Tons per Capita
U.S.	5,475	20.52
China	3,196	2.68
Russian Federation	1,820	12.26
Japan	1,126	9.03
India	910	0.90
Germany	833	10.24
U.K.	539	9.29
Ukraine	437	8.48
Canada	433	14.83
Italy	411	7.19
South Korea	370	8.33
Mexico	359	3.93

John J. Fialka, *Wall Street Journal*, October 3, 1997.

The Clinton administration has made trading a main part of its negotiating position on the treaty to prevent global warming. The treaty would impose a global limit on man-made sources of carbon dioxide, created by burning petroleum products, coal and natural gas. Since the 1850s, concentrations of carbon dioxide have increased in the atmosphere, and many scientists say that is artificially warming the Earth by trapping more of the sun's heat.

Draft versions of the U.S. plan would give each industrial nation an "emissions budget" that would be its pro rata share of the global limit. (A program for developing nations would be negotiated at a later time.) Industrial nations that curb emissions below 1990 levels would have allowances to sell. Countries that can't curb it, or won't, must buy them.

Adapting the U.S. acid-rain program to do that means stretching a trading program designed to cover hundreds of emitters to one that would cover millions and allowing international transactions. This will compound the complexity of trading, but Mr.

Dudek insists it will also create many more cheap opportunities to reduce global carbon-dioxide levels.

For example, a U.S. company could buy more efficient boilers or expensive antipollution equipment to meet its share of U.S. carbon-dioxide-emissions quota. Or, it could invest in more cost-effective clean-up projects in Russia. In a global program, Russia will have a great many allowances to sell because its industrial economy has collapsed since 1990, sharply cutting its emissions. And 1990 is expected to be the benchmark year for the treaty.

According to Australian analysts, the U.S., New Zealand, Japan, Norway and their country will have to buy emission allowances to meet their limits. European nations will, more or less, balance each other out. So the world's excess certificates would be held by Russia and the rest of the former Soviet empire.

"We're talking about transfers of billions of dollars here," says Stephen Deady, commercial minister at Australia's embassy in Washington. "We think the whole question of emissions trading needs further work." European nations say they will consider trading plans, but only after the Clinton administration announces how much it will reduce U.S. emissions.

Japan, anxious for meaningful results, has requested more details of the U.S. position. China, expected to be a major player in the Kyoto negotiations as a leader of developing nations, may be of two minds. Chinese diplomats reject carbon-dioxide controls as "ecocolonialism." But some officials see trading as an additional means of getting outside investors to clean up most of China's severe pollution problems in exchange for trading certificates.

Despite the faint praise from its allies, U.S. officials remain bullish. Although the Russians haven't publicly stated their position, Energy Secretary Federico Pena says Russian officials are "very interested" in emissions trading. To sell certificates, he explains, Russia would first have to set up a rigorous monitoring system so the money from the sales could be focused on cleaning up the country's dirtiest factories or its enormous natural-gas-distribution system, which is believed to be leaking huge quantities of methane, a more potent greenhouse gas than carbon dioxide.

Mr. Pena sees a situation in which "a U.S. company could go over, make a major investment to reduce the leakage and get credit for that. That would be a win for the local people, the government, the U.S. company and a win for global warming."

But the idea baffles some groups. "When this gets to the real world, there's no way it can work," says Daniel Becker, director of the Sierra Club's energy program. "You're going to let Russia be able to sell credit for plants that have already been closed so that General Motors can emit more pollution here?"

"WIN-WIN" DEALS

The Clinton administration says trading also offers "win-win" deals for developing nations. The Energy Department has been helping Costa Rica develop a program based on reforestation. For outside investors, Costa Rica will agree to protect a portion of its rain forest that would otherwise be logged. Because standing forests remove carbon dioxide from the air, Costa Rica will then issue trading certificates.

Richard L. Sandor, second vice chairman of the Chicago Board of Trade, has bought 1,000 of them, each permitting the owner to emit 1,000 tons of carbon dioxide. He hopes to sell them to U.S. companies in anticipation of a global-trading plan. "If Kyoto gives us a framework, that's all we need," says Mr. Sandor, who has been studying the possibility of carbon-dioxide trading for the Board of Trade. He says a world-wide market could generate $10 billion to $20 billion a year, rivaling the size of the U.S. soybean market.

But Mr. Sandor will have a hard time selling any of this to Bill Hare, a physicist and global-warming expert for Greenpeace International in Amsterdam. Costa Rica's effort "worries the hell out of us," Mr. Hare says. He explains that there is no international-accounting system to ensure that the trees will remain standing.

Still, he says some kind of trading regime could come out of Kyoto, though it may take years to negotiate the complex rules. Swarms of investment bankers and venture capitalists are showing up at global-warming meetings, once ill-attended affairs dominated by a few scientists. "This is a very interesting development," he says.

Meanwhile traders continue to push the envelope. As the result of a modernization program, New York state's Niagara Mohawk Power Corp. cut its carbon-dioxide emissions by 2.5 million tons. It traded emission rights for that amount to an Arizona utility for 20,000 tons of sulfur-dioxide allowances. Then Niagara donated the sulfur-dioxide allowances to local environmental groups, which are retiring them. And $125,000 of the tax benefits Niagara received from the deal are being used to convert a Mexican fishing village to nonpolluting solar power.

Martin A. Smith, chief environmental scientist at Niagara, says the reductions and the experience in trading were worth it. "Do we just stand by and let these gases accumulate?"

> "Air quality can continue to decline, as companies in some parts of the country simply buy their way out of pollution reductions."

POLLUTION CREDITS WILL INCREASE AIR POLLUTION

Brian Tokar

The pollution credit program that was passed in 1990 as an amendment to the Clean Air Act has done nothing to reduce pollution except to lower companies' costs in complying with the reduced emissions levels, contends Brian Tokar in the following viewpoint. Pollution credits simply give companies the "right" to pollute, he argues. Technological improvements, not pollution credits, are responsible for the declining emissions levels, Tokar maintains. Tokar is the author of *Earth for Sale* and *The Green Alternative*. He teaches at the Institute for Social Equality and at Goddard College, both in Plainfield, Vermont.

As you read, consider the following questions:

1. How much were pollution credits originally expected to sell for, and what did they actually sell for in 1993, according to the author?
2. How do some companies reduce their pollution emissions without buying credits, as cited by Tokar?
3. What does the author predict will happen if pollution credits are allowed to be bought and sold like any other commodity?

Excerpted from Brian Tokar, "Trading Away the Earth," *Dollars & Sense*, March/April 1996. Reprinted with permission. *Dollars & Sense* is a progressive economics magazine published six times a year. First-year subscriptions cost $18.95 and may be ordered by writing to *Dollars & Sense*, One Summer St., Somerville, MA 02143.

The Republican takeover of Congress has unleashed an un-precedented assault on all forms of environmental regulation. From the Endangered Species Act to the Clean Water Act and the Superfund for toxic waste cleanup, laws that may need to be strengthened and expanded to meet the environmental challenges of the next century are instead being targeted for complete evisceration.

FREE MARKET ENVIRONMENTALISM

For some activists, this is a time to renew the grassroots focus of environmental activism, even to adopt a more aggressively anti-corporate approach that exposes the political and ideological agendas underlying the current backlash. But for many, the current impasse suggests that the movement must adapt to the dominant ideological currents of the time. Some environmentalists have thus shifted their focus toward voluntary programs, economic incentives and the mechanisms of the "free market" as means to advance the cause of environmental protection. Among the most controversial, and widespread, of these proposals are tradeable credits for the right to emit pollutants. These became enshrined in national legislation in 1990 with President George Bush's amendments to the 1970 Clean Air Act.

Even in 1990, "free market environmentalism" was not a new phenomenon. In the closing years of the 1980s, an odd alliance had developed among corporate public relations departments, conservative think tanks such as the American Enterprise Institute, Bill Clinton's Democratic Leadership Council (DLC), and mainstream environmental groups such as the Environmental Defense Fund. The market-oriented environmental policies promoted by this eclectic coalition have received little public attention, but have nonetheless significantly influenced debates over national policy.

Glossy catalogs of "environmental products," television commercials featuring environmental themes, and high profile initiatives to give corporate officials a "greener" image are the hallmarks of corporate environmentalism in the 1990s. But the new market environmentalism goes much further than these showcase efforts. It represents a wholesale effort to recast environmental protection based on a model of commercial transactions within the marketplace. "A new environmentalism has emerged," writes economist Robert Stavins, who has been associated with both the Environmental Defense Fund and the DLC's Progressive Policy Institute, "that embraces . . . market-oriented environmental protection policies."

Today, aided by the anti-regulatory climate in Congress, market schemes such as trading pollution credits are granting corporations new ways to circumvent environmental concerns, even as the same firms try to pose as champions of the environment. While tradeable credits are sometimes presented as a solution to environmental problems, in reality they do nothing to reduce pollution—at best they help businesses reduce the costs of complying with limits on toxic emissions. Ultimately, such schemes abdicate control over critical environmental decisions to the very same corporations that are responsible for the greatest environmental abuses.

How It Works, and Doesn't

A close look at the scheme for nationwide emissions trading reveals a particular cleverness; for true believers in the invisible hand of the market, it may seem positively ingenious. Here is how it works: The 1990 Clean Air Act amendments were designed to halt the spread of acid rain, which has threatened lakes, rivers and forests across the country. The amendments required a reduction in the total sulfur dioxide emissions from fossil fuel burning power plants, from 19 to just under 9 million tons per year by the year 2000. These facilities were targeted as the largest contributors to acid rain, and participation by other industries remains optional. To achieve this relatively modest goal for pollution reduction, utilities were granted transferable allowances to emit sulfur dioxide in proportion to their current emissions. For the first time, the ability of companies to buy and sell the "right" to pollute was enshrined in U.S. law.

Any facility that continued to pollute more than its allocated amount (roughly half of its 1990 rate) would then have to buy allowances from someone who is polluting less. The 110 most polluting facilities (mostly coal burners) were given five years to comply, while all the others would have until the year 2000. Emissions allowances were expected to begin selling for around $500 per ton of sulfur dioxide, and have a theoretical ceiling of $2000 per ton, which is the legal penalty for violating the new rules. Companies that could reduce emissions for less than their credits are worth would be able to sell them at a profit, while those that lag behind would have to keep buying credits at a steadily rising price. For example, before pollution trading every company had to comply with environmental regulations, even if it cost one firm twice as much as another to do so. Under the new system, a firm could instead choose to exceed the mandated levels, purchasing credits from the second firm instead of implementing costly controls. This exchange would save money, but in principle yield the same overall level of pollution as if both companies had complied equally. Thus, it is argued, market forces will assure that the most cost-effective means of reducing acid rain will be implemented first, saving the economy billions of dollars in "excess" pollution control costs.

INVESTMENT AND INNOVATION

Defenders of the Bush plan claimed that the ability to profit from pollution credits would encourage companies to invest more in new environmental technologies than before. Innovation in environmental technology, they argued, was being stifled by regulations mandating specific pollution control methods. With the added flexibility of tradeable credits, companies could postpone costly controls—through the purchase of some other company's credits—until new technologies became available. Proponents argued that, as pollution standards are tightened over time, the credits would become more valuable and their owners could reap large profits while fighting pollution.

Yet the program also included many pages of rules for extensions and substitutions. The plan eliminated requirements for backup systems on smokestack scrubbers, and then eased the rules for estimating how much pollution is emitted when monitoring systems fail. With reduced emissions now a marketable commodity, the range of possible abuses may grow considerably, as utilities will have a direct financial incentive to manipulate reporting of their emissions to improve their position in the pollution credits market.

NOT ACCORDING TO PLAN

Once the EPA actually began auctioning pollution credits in 1993, it became clear that virtually nothing was going according to their projections. The first pollution credits sold for between $122 and $310, significantly less than the agency's estimated minimum price, and by 1995, bids at the EPA's annual auction of sulfur dioxide allowances averaged around $130 per ton of emissions. As an artificial mechanism superimposed on existing regulatory structures, emissions allowances have failed to reflect the true cost of pollution controls. So, as the value of the credits has fallen, it has become increasingly attractive to buy credits rather than invest in pollution controls. And, in problem areas air quality can continue to decline, as companies in some parts of the country simply buy their way out of pollution reductions.

At least one company has tried to cash in on the confusion by assembling packages of "multi-year streams of pollution rights" specifically designed to defer or supplant purchases of new pollution control technologies. "What a scrubber really is, is a decision to buy a 30-year stream of allowances," John B. Henry of Clean Air Capital Markets told the *New York Times*, with impeccable financial logic. "If the price of allowances declines in future years," paraphrased the *Times*, "the scrubber would look like a bad buy."

Reprinted by permission of Chip Bok and Creators Syndicate.

Where pollution credits have been traded between companies, the results have often run counter to the program's stated intentions. One of the first highly publicized deals was a sale of credits by the Long Island Lighting Company to an unidentified company located in the Midwest, where much of the pollution that causes acid rain originates. This raised concerns that places suffering from the effects of acid rain were shifting "pollution rights" to the very region it was coming from. One of the first companies to bid for additional credits, the Illinois Power Company, canceled construction of a $350 million scrubber system in the city of Decatur, Illinois. "Our compliance plan is based almost totally on purchase of credits," an Illinois Power spokesperson told the *Wall Street Journal*. The comparison with more traditional forms of commodity trading came full circle in 1991, when the government announced that the entire system for trading and auctioning emissions allowances would be administered by the Chicago Board of Trade, long famous for its ever-frantic markets in everything from grain futures and pork bellies to foreign currencies.

OTHER STRATEGIES

Some companies have chosen not to engage in trading pollution credits, proceeding with pollution control projects, such as the installation of new scrubbers, that were planned before the credits became available. Others have switched to low-sulfur coal and increased their use of natural gas. If the 1990 Clean Air Act amendments are to be credited for any overall improvement in air quality, it is clearly the result of these efforts and not the market in tradeable allowances.

Yet while some firms opt not to purchase the credits, others, most notably North Carolina–based Duke Power, are aggressively buying allowances. At the 1995 EPA auction, Duke Power alone bought 35% of the short-term "spot" allowances for sulfur dioxide emissions, and 60% of the long-term allowances redeemable in the years 2001 and 2002. Seven companies, including five utilities and two brokerage firms, bought 97% of the short-term allowances that were auctioned in 1995, and 92% of the longer-term allowances, which are redeemable in 2001 and 2002. This gives these companies significant leverage over the future shape of the allowances market.

The remaining credits were purchased by a wide variety of people and organizations, including some who sincerely wished to take pollution allowances out of circulation. Students at several law schools raised hundreds of dollars, and a group at the Glens Falls Middle School on Long Island raised $3,171 to purchase 21 allowances, equivalent to 21 tons of sulfur dioxide emissions over the course of a year. Unfortunately, this represented less than a tenth of one percent of the allowances auctioned off in 1995.

Some of these trends were predicted at the outset. "With a tradeable permit system, technological improvement will normally result in lower control costs and falling permit prices, rather than declining emissions levels," wrote Robert Stavins and Brad Whitehead (a Cleveland-based management consultant with ties to the Rockefeller Foundation) in a 1992 policy paper published by the Progressive Policy Institute. Despite their belief that market-based environmental policies "lead automatically to the cost-effective allocation of the pollution control burden among firms," they are quite willing to concede that a tradeable permit system will not in itself reduce pollution. As the actual pollution levels still need to be set by some form of regulatory mandate, the market in tradeable allowances merely gives some companies greater leverage over how pollution standards are to be implemented.

Without admitting the underlying irrationality of a futures market in pollution, Stavins and Whitehead do acknowledge (albeit in a footnote to an Appendix) that the system can quite easily be compromised by large companies' "strategic behavior." Control of 10% of the market, they suggest, might be enough to allow firms to engage in "price-setting behavior," a goal apparently sought by companies such as Duke Power. To the rest of us, it should be clear that if pollution credits are like any other commodity that can be bought, sold and traded, then the largest "players" will have substantial control over the entire "game." Emissions trading becomes yet another way to assure that large corporate interests will remain free to threaten public health and ecological survival in their unchallenged pursuit of profit.

| "A gasoline car will over the course of its working life emit twice as much nitrogen oxides (NOx) as will a generating facility producing the power to charge a comparable electric vehicle."

ELECTRIC CARS WILL REDUCE AIR POLLUTION

Drew Kodjak

In the following viewpoint, Drew Kodjak contends that electric vehicles (EVs) are much cleaner than gasoline-powered automobiles and better for the environment. Gasoline vehicles emit more pollutants from their exhaust and during refueling than do EVs, as do the oil refineries and ships and trucks that transport the fuel. Cleaner air is an immediate benefit with electric cars, Kodjak asserts, while the benefits of more stringent regulations and cleaner gasoline vehicles must wait until the older, dirtier vehicles are replaced. Kodjak is an attorney for mobile sources at Northeast States for Coordinated Air Use Management in Boston, an interstate association of air quality control agencies.

As you read, consider the following questions:

1. According to EPA statistics, what is the increase in tailpipe emissions from gasoline vehicles as mileage accumulates?
2. How many years does it take before 95 percent of the fleet of gasoline-powered automobiles is taken off the roads, according to Kodjak?
3. What are some alternative methods suggested by the author for powering electric vehicles?

Reprinted with permission from *Technology Review*, published by the Association of Alumni and Alumnae of MIT, copyright 1998. "EVs: Clean Today, Cleaner Tomorrow," by Drew Kodjak, August/September 1996. Reproduced by permission of the publisher via Copyright Clearance Center, Inc.

The debate about electric vehicles (EVs) often revolves around whether increased smokestack emissions—from the power plants generating the electricity needed to recharge all those EV batteries—will offset the reduction in pollution from tailpipes. Such analyses usually overlook a critical factor, however: gasoline cars tend to get dirtier and dirtier over time, while electric power plants do not.

CARS BECOME DIRTIER

According to the U.S. Environmental Protection Agency, pollution from tailpipes grows by an average of 25 percent every 10,000 miles, culminating in vehicles that are 2 to 5 and sometimes 10 times dirtier than when they left the assembly line. Indeed, because of this steadily deteriorating efficiency, a gasoline car will over the course of its working life emit twice as much nitrogen oxides (NOx) as will a generating facility producing the power to charge a comparable electric vehicle. The conventional car will also spew out 60 times more carbon monoxide, 30 times more volatile organic compounds, and twice the carbon dioxide emissions as the electric power plant. This comparison assumes that electric vehicles will be recharged with power drawn mainly from oil- and natural gas–fired generators.

Maintaining the emissions control systems in millions of individually owned vehicles has proved extremely difficult. Indeed, in the northeastern United States, the single largest source of air pollution is aging gasoline-powered vehicles. Recognizing this fact, several Northeast states announced plans in 1993 to upgrade their automobile inspection and maintenance systems. In some states, these programs are expected to eventually achieve up to 40 percent of the reductions in volatile organic compound emissions necessary to comply with new clean air standards. Unfortunately, most of these programs have been delayed, in large part due to concern about public response to the cost of repairing catalytic converters and other such components. Utilities, by contrast, employ teams of professional engineers to keep the power plants well maintained and operating at peak efficiency.

AN IMMEDIATE REDUCTION IN AIR POLLUTION

Another benefit unique to electric vehicles is the immediate reduction in air pollution over the entire EV fleet once new power-plant-emission controls are installed. Under pressure from regulators, for example, electric utilities in the Northeast have committed to reducing power plant emissions of NOx by

55 to 75 percent over the next seven years. The entire fleet of EVs will reap the environmental advantage of this upgrade at a few hundred facilities. With gasoline vehicles, by contrast, the benefits from more stringent new emissions standards are realized only incrementally as older vehicles are junked and replaced. Since 95 percent of the fleet turns over only after 12–15 years, more than a decade can elapse before the desired emissions reductions are achieved.

Because of the rising popularity of gas-guzzling sport utility vehicles, which now account for more than 40 percent of new vehicle sales, the overall efficiency of passenger cars has dropped in the last several years. EVs, on the other hand, are entering the market as highly fuel-efficient machines. The General Motors EV1, for example, gets the equivalent of 100 miles per gallon.

Moreover, gasoline vehicles of all kinds are often restarted several times a day, and, once running, the engine must frequently change speed as the vehicle accelerates, decelerates, and traverses hills. Such sporadic and variable operation works against fuel efficiency. EVs are charged by electric generators that reap the efficiency benefit of running continuously at a constant speed.

A Cleaner Way to Travel

The fact is, every study that's been done shows it's at least 95 percent less polluting to drive an electric car versus an internal-combustion engine. Do you realize that you use 25 percent of your gasoline in a big city *stopped* in traffic? Now I'm on my third electric vehicle, a converted VW Rabbit that runs like a top. . . .

Ninety percent of our trips are 40 miles or less. Well, even the most lame-ass electric car can go that far without a recharge.

Ed Begley Jr., interviewed by Dick Russell, E *Magazine*, January/February 1996.

Not all automobile pollution comes from the tailpipe. About a third of the volatile organic compounds that gasoline automobiles introduce into the atmosphere evaporate during refueling and from the gas tank, engine, and fuel line. Significant additional emissions come from the operation of refineries and from ships and trucks that transport the liquid fuel. Electric vehicles reduce these emissions as well as exposure at the gas station to the toxic chemicals present in gasoline, such as benzene and 1,3-butadiene.

Over the next fifteen years, EV batteries in the Northeast will increasingly be recharged by modern natural gas–fired power plants. These plants are much cleaner and more efficient than today's coal or oil-fired plants. For example, facilities in Southern California already emit 40 times less NOx than Northeast power plants.

OTHER POWER SOURCES

And unlike conventional vehicles, which run almost exclusively on petroleum-based fuels, electric vehicles are able to tap into a large number of power sources, including renewables such as hydro, wind, geothermal, and biomass. In a demonstration project run by the Massachusetts energy office, solar cells provide the electricity to recharge EVs parked at two commuter rail train stations in the Boston area. Solar panels on the roofs of houses could collect solar energy by day and use it to charge a spare EV battery. Once home, the motorist would simply exchange today's spent battery pack with the newly charged one.

The move to EVs must be understood as a long-range air-quality strategy. It will take many years from the time EVs are introduced until even half the cars on the road are electric. Ultimately, EVs can make a difference only if people buy them. In the near term, consumers will have to choose between internal-combustion vehicles that offer the romantic roar of a V8 and 500-mile road trips, and electric vehicles that zoom silently from 0 to 60 and can be recharged overnight in their garage. Over the long term the clean-air payback will become apparent as gracefully aging EVs supplant the elderly, atmospherically incontinent fleet of gasoline cars.

"*Because producing electricity also creates pollution, electric vehicles do not eliminate emissions—they simply move them elsewhere.*"

ELECTRIC CARS WILL NOT REDUCE AIR POLLUTION

Richard de Neufville et al.

In the following viewpoint, Richard de Neufville, Stephen R. Connors, Frank R. Field III, David Marks, Donald R. Sadoway, and Richard D. Tabors argue that requiring car manufacturers to build electric vehicles instead of conventional automobiles to reduce air pollution will be costly and ineffective. De Neufville and his colleagues maintain that the electricity produced to power electric vehicles also creates pollution; therefore, electric vehicles do not reduce air pollution but merely move the emissions to another location. Due to continuing improvements in reducing emissions in conventional vehicles, electric vehicles will have a negligible effect on reducing air pollution, they assert. Instead of mandating electric vehicles as a means of reducing air pollution, the authors recommend more research and development of alternative vehicles. The authors are all based at the Massachusetts Institute of Technology in various engineering, technology, and energy departments.

As you read, consider the following questions:

1. What percentage of cars sold in 2003 must be electric-powered, according to the authors?
2. What are the environmental problems with the three main types of power plants that could provide electricity for electric cars, in the authors' opinion?

To comply with the federal Clean Air Act of 1990, the California Air Resources Board has ruled that, by 1998, 2 percent of all vehicles offered for sale in the state must be so-called zero-emission vehicles. As a practical matter, California has mandated electric vehicles—the only available technology meeting the requirement that the power train produce no emissions. Two other states have followed suit. By 2003, roughly 10 percent of all new personal vehicles sold in California, Massachusetts, and New York must be electric.

The aim of these programs—to combat the smog that engulfs Los Angeles and other cities—is worthy. Even the programs' focus on cars is appropriate because there is little question that auto emissions contribute greatly to urban air pollution. Unfortunately, however, the electric vehicle is not yet ready for large-scale commercial use. No such vehicle now sold meets the demands of a consumer market for road transport.

Highway-worthy electric vehicles for mass consumption have neither been produced nor tested in significant volumes over the range of likely driving conditions. Their reliability over a standard warranty period, such as 3 years and 50,000 miles, is unknown. Electric vehicles for actual road use are still highly experimental.

CAREFUL SCRUTINY IS NEEDED

The mandate to produce and sell a significant number of electric vehicles thus needs careful scrutiny. The measure is unprecedented. Previous environmental mandates, such as the Clean Water Act, required the public to adopt the best available technology—whatever that turned out to be in different cases—for reducing pollution. The California rules, however, require a specific experimental technology, and mandate a tight schedule.

The effort to pursue electric vehicles on a large scale is also uniquely American. Britain uses electric vehicles for milk delivery, France has proposed pilot production of special urban vehicles, and the German Post Office wants to operate about a hundred delivery vans in the years ahead. In addition, Volkswagen has recently started to produce electric-powered Golf sedans at the rate of about one per day. But despite this interest in electric vehicles, the existing programs in other countries are orders of magnitude smaller than what is required by the California rules, which aim for manufacturers to sell some 18,000 electric vehicles in 1998 (some 2 percent of the 906,000 new cars registered in California in 1994).

Meanwhile, in the United States, the program to develop electric vehicles has already proved expensive. Ford and General Motors alone have reportedly spent hundreds of millions of dollars on R&D, and federal and state agencies have sponsored a wide range of demonstration programs. The budget for the U.S. Advanced Battery Consortium alone (an alliance of the U.S. Department of Energy, the Big Three automobile manufacturers, the Electric Power Research Institute, Southern California Edison, and others to develop batteries for the vehicles) is $260 million. In fact, the cumulative cost of research on electric vehicles in the United States is approaching $1 billion, roughly equal to half of the National Science Foundation's entire research budget.

AN INVESTIGATION

By any measure then, the commitment to manufacture and sell electric vehicles in large volume is a major piece of national industrial policy that aims to substantially reduce the nation's transportation and pollution problems. One supposes that such a mandate would have been preceded by a comprehensive analysis. Yet no investigation of the overall performance or effectiveness of electric vehicles—either by themselves or compared with alternatives—has been undertaken. Our research group found that available material either deals with just one element of the system, such as batteries, or is obviously partisan, coming from enthusiasts—such as electric vehicle makers, battery suppliers, or electric utilities—with a stake in the outcome.

To address this gap, our team assessed the total environmental and economic effects of the manufacture and use of electric vehicles made with different materials and powered by many types of batteries. We also attempted to compare the electric-car mandate to alternative systems for reducing air pollution.

In our judgment, the electric vehicle policy defined by the California Air Resources Board is neither cost-effective nor practical. Electric vehicles will not contribute meaningfully to cleaner air if they are introduced as now proposed; over the next decade their effect will be imperceptible compared with other major improvements in automotive and other combustion technologies. Furthermore, even if it could be justified on environmental grounds, the technology of electric vehicles is still far from meeting the needs of a mass consumer market and it is unclear when, if ever, it will do so. Finally, the projected costs of implementing the California electric vehicle policy are enormous, requiring subsidies as high as $10,000 to $20,000 per vehicle.

Because conventional cars and trucks create significant emissions, the use of electric vehicles sounds like a good way to combat air pollution. But because producing electricity also creates pollution, electric vehicles do not eliminate emissions— they simply move them elsewhere. Unless this electricity comes from nuclear power plants (neither environmentally acceptable nor economically feasible right now) or renewable sources (unlikely to be sufficient), the power to propel electric vehicles will come from burning fossil fuels. But using fossil fuels to power electric vehicles is doubly pernicious. The fuel loses up to 65 percent of its energy when it is burned to produce electricity; 5 to 10 percent of what is left is lost in transmitting and distributing the electricity before it even gets to the electric car.

Of course, moving pollutant emissions elsewhere could arguably be worthwhile, but such a policy needs to be considered carefully. For regions upwind of power plants, electric vehicles would obviously reduce local pollution. Los Angeles, for instance, obtains part of its electric power from coal plants in the Four Corners region (where Arizona, Colorado, New Mexico, and Utah meet). Adopting electric vehicles in Los Angeles therefore simply increases pollution over large expanses of the Southwest.

LITTLE REDUCTION IN TAILPIPE EMISSIONS

Consumer Reports, long an advocate of cleaner cars, cited a researcher from the University of California Institute of Transportation Studies who calculated that replacing *all* gasoline-burning cars in the United States with electric ones would reduce tailpipe emissions by only 20 percent. Said the researcher, "The same improvement could be achieved at lower cost just by improving the efficiency of gas-burning cars."

Ralph Kinney Bennet, *Reader's Digest*, January 1998.

Meanwhile, regions downwind of fossil fuel–burning power plants, such as Boston and the Northeast seaboard generally, will not escape the pollution produced by generating electricity for electric vehicles, which may be substantial. What's more, many discussions of electric vehicles have supposed that the plants used to create the extra power would be clean and inexpensive, since the electric cars would mostly be recharged "off peak." But this is unlikely to be the case. Much of the power from the cleanest and least expensive plants is already in use today even during off-peak hours; supplying the additional loads will inevitably require using older, dirtier, and less efficient facilities.

Even in areas where electric cars may lower urban air pollution, the great effort to get them on the road may not perceptibly improve the environment. For the past decade, new vehicles have met more stringent pollution standards, and the upgrading of the fleet has cut total U.S. automotive emissions dramatically. Even without electric vehicles, the fleet of cars now on the road will be almost completely renewed in this decade and thus the average emissions from cars will be almost halved. Ironically, the environmental benefit of each electric vehicle would be particularly small in the years ahead because it would substitute for another brand-new vehicle that will be far cleaner than the current average.

The schedule for the introduction of new electric vehicles implies that only about 4 percent of the total fleet in California, Massachusetts, and New York will be electric by the year 2005, and about 10 percent by the year 2015, some 20 years from now. And improvement will not be immediate: since only about 10 percent of the automotive fleet is renewed each year in the United States, it takes about a decade for the percentage of electric vehicles on the road to match the percentage of those sold each year. Thus given the small percentages involved and the long delays, electric vehicles will have only a modest effect on overall automotive pollution. This is true of any policy that imposes marginal improvements on a small fraction of the cars on the road. The important effects result from changes to the entire fleet. Thus the requirement that all cars use catalytic converters to limit carbon monoxide emissions improved air quality significantly, but the California mandate to introduce electric vehicles will not. . . .

DRIVING FORWARD

Unfortunately, because today's policy fixates prematurely on a specific technological solution, it has diverted attention from the basic issue: How should we improve air quality in polluted urban areas? To obtain a practical result, we need to consider both the instrument of the problem—that is, the technology—and the cause of the problem, the users. We need to adopt a flexible strategy that permits us to choose the most effective options as they develop. We must also define approaches that can command the support of all the important participants.

Rather than mandate development of the electric vehicle on a short timetable, we should promote research and development over a broad front on a range of alternative vehicles. These should certainly include refined versions of currently accessible technologies such as ultra-low-emission vehicles that use catalytic converters and microelectronics to control combustion precisely; and so-called hybrid vehicles, which combine constant-speed (and therefore highly efficient) gasoline or diesel engines with electric generators to extend the range and power of batteries stored on board. Fuel-cell vehicles are a technological possibility that also requires investigation.

Development could also be divided into three phases. The first might focus on creating prototypes, culminating in a competition between technologies. The second phase could then concentrate on large-scale development and testing of finalist systems, leading to a final choice for implementing in the third phase. In light of all the uncertainties, it is unlikely that a particular schedule for such implementation, set a decade in advance, can work.

Organizational changes should also complement, or even replace, technological solutions. Perhaps the real issue is that communities such as Los Angeles are too dependent on the use of personal automobiles. Because the total level of pollution is of course the product of two factors—the dirtiness of the vehicle and the distance it travels—targeting the level of emissions produced per vehicle-mile addresses only half the problem. The fact is that the number of vehicle-miles traveled is growing steadily in the United States, particularly in the Los Angeles area. More people live farther away from jobs and travel more. If this trend continues, the resulting increase in pollution will counteract any reduction achieved by introducing electric vehicles. An effective policy to reduce total automotive pollution should thus include encouraging collective transport through the use of car pools and buses, reducing driving through disincentives such as higher parking fees and gas taxes, and facilitating alternatives to driving such as telecommuting.

CHEAP AND IMMEDIATE ACTIONS

In the first phase of any such plan, decision makers should identify actions that can produce immediate results cheaply—in essence, picking the low-hanging fruit. They should, for example, consider a program of buying up the most severely polluting vehicles—those among the 7 to 10 percent of vehicles that produce 50 percent of on-road generation of carbon monoxide and hydrocarbons. Because one of these mostly older, severely polluting vehicles produces roughly 10 times the pollution of an average vehicle, and because one electric vehicle will only reduce pollution equal to one-half of an average car, such a program would have 20 times the effect per vehicle and would be far more cost-effective.

Such a multifaceted and dynamic strategy would surely improve air quality more quickly than a proposed mandate that will have no perceptible effect on pollution for many years, if ever.

Periodical Bibliography

The following articles have been selected to supplement the diverse views presented in this chapter. Addresses are provided for periodicals not indexed in the *Readers' Guide to Periodical Literature*, the *Alternative Press Index*, the *Social Sciences Index*, or the *Index to Legal Periodicals and Books*.

E. Calvin Beisner	"How Environmentalism Disdains the Poor," *Freeman*, August 1998. Available from the Foundation for Economic Education, Inc., Irvington-on-Hudson, NY 10533.
Ralph Kinney Bennet	"Electric Cars? Fahgedaboudit!" *Reader's Digest*, January 1998.
Mary H. Cooper	"New Air Quality Standards," *CQ Researcher*, March 7, 1997. Available from 1414 22nd St. NW, Washington, DC 20037.
Robert W. Crandall	"The Costly Pursuit of the Impossible," *Brookings Review*, Summer 1997.
John H. Cushman Jr.	"EPA and States Found to Be Lax on Pollution Law," *New York Times*, June 7, 1998.
Terry Davies	"Critically Evaluating America's Pollution Control System," *Resources*, Winter 1998. Available from 1616 P St. NW, Washington, DC 20036-1400.
Brian Doherty	"Selling Air Pollution," *Reason*, May 1996.
James M. Inhofe	"Federal Regulators Are Set to Choke Local Economies for the Sake of Marginal Health Benefits," *Insight*, April 28, 1997. Available from 3600 New York Ave. NE, Washington, DC 20002.
Lester B. Lave	"Clean Air Sense," *Brookings Review*, Summer 1997.
Laura M. Litvan	"A Breath of Fresh Air," *Nation's Business*, May 1995.
John Semmens	"The Environmental Assault on Mobility," *Freeman*, August 1995.
Daniel Sperling	"The Case for Electric Vehicles," *Scientific American*, November 1996.

CHAPTER 5

HOW SHOULD POLLUTION BE MANAGED?

CHAPTER PREFACE

During the late 1800s, William T. Love started digging a canal near Niagara Falls, New York, but never finished it due to an economic depression. In the early 1940s, Hooker Chemical Company bought the property encompassing the 3,000-foot trench and began to use it as a chemical waste dump. Meanwhile, the city of Niagara Falls was expanding rapidly and the city began exerting enormous pressure on Hooker Chemical Company to sell the Love Canal for development. Hooker Chemical resisted until 1953, when the city threatened to condemn the property to force the sale. At that point, Hooker published reports in the local newspaper outlining the potential dangers of the site due to the hazardous wastes buried in the canal and emphasized that no excavating should be done on the site. Hooker then sold the land to the city for $1. The city went forward with plans to build an elementary school on a portion of the dump and sold the rest of the site to a developer for residential housing.

During construction of the school and houses, the waste dump's lining was broken, allowing hazardous chemicals to seep into the soil and water. The leaking went undetected until 1974, when the dump's surface collapsed, exposing barrels of chemicals. In 1977 a local newspaper reporter wrote articles alleging links between the hazardous wastes buried in Love Canal and various illnesses suffered by Love Canal residents. The story received national press and led to passage of the Comprehensive Environmental Response, Compensation, and Liability Act of 1980, more commonly known as Superfund.

Under Superfund, $15 billion dollars from taxes on polluting industries was to be used to clean up the 400 most polluted hazardous waste sites in the country. As of February 1998, the Environmental Protection Agency has expanded that list to 1,191 contaminated sites. Some analysts predict that $500 billion in cleanup costs will have been spent on the Superfund program by 2030. The state of New York, the federal government, and Love Canal residents all sued Hooker Chemical's successor, Occidental Chemical, which was forced to pay $98 million to New York and $129 million to the federal government for cleanup costs, and between $83 and $400,000 to 2,300 Love Canal families. Despite Love Canal's history, new homes have been built on the former waste dump and there is a waiting list of families for renovated homes.

The viewpoints in the following chapter examine the Superfund program and other forms of pollution control, and discuss who should be in charge of these programs.

> "Despite expenditures of between $20 billion and $30 billion, Superfund has failed to clean up more than a small fraction of the nation's worst hazardous waste sites."

THE STATES SHOULD MANAGE THE SUPERFUND PROGRAM

John Shanahan

Superfund was originally conceived of as a temporary program to clean up a few hazardous waste sites across the United States. Instead, argues John Shanahan in the following viewpoint, Superfund has become an ever-expanding permanent federal program that wastes billions of dollars in an impossible attempt to return contaminated sites that pose little danger to people to a pristine condition. Shanahan contends that, since contaminated sites are a local, and not a national, problem, the Superfund program should be reformed to give the states the authority to clean up the hazardous waste sites. Shanahan is a policy analyst with the Heritage Foundation, a conservative public policy organization in Washington, D.C.

As you read, consider the following questions:

1. What are the four fundamental flaws of the Superfund program, in Shanahan's opinion?
2. According to the author, why has the Superfund had little effect on public health?
3. How does the Superfund liability scheme encourage legal gridlock, cleanup delay, and enormous legal costs, according to Shanahan?

Excerpted from John Shanahan, "How to Rescue Superfund: Bringing Common Sense to the Process," *Backgrounder*, no. 1047, July 31, 1995. Reprinted by permission of The Heritage Foundation.

The 1980 Comprehensive Environmental Response, Compensation, and Liability Act (CERCLA) [or] Superfund, as CERCLA is popularly known, not only is one of the most complex laws now in force, but also is widely regarded as both wasteful and ineffective. Even the Clinton Administration admits the program does not work. One reason for this is that, despite expenditures of between $20 billion and $30 billion, Superfund has failed to clean up more than a small fraction of the nation's worst hazardous waste sites. Under the current law, cleanup of the remaining current and potential sites could cost additional tens of billions of dollars. . . .

Congress must correct the fundamental flaws in the current program, not merely tinker with the statute. . . . Reform should address the core elements of the Superfund structure that fuel litigation, slow cleanups, raise costs, and, most important, encourage the cleanup of sites that pose little risk to human health and safety and divert scarce economic resources from sites that may pose real dangers. . . .

LARGE COSTS AND FEW RESULTS

The Superfund program was enacted after widespread press attention to the exposure of residents to high levels of dangerous chemicals at Love Canal near Niagara Falls, New York. Intended to force companies to clean up hazardous waste sites, the law also established a government fund to pay for cleanup of "orphan sites," which are contaminated properties where no responsible party exists—where the polluter, for example, has gone out of business. The law was hurried through Congress in the final days of the Carter Administration, and many provisions were ill-considered. Although estimates vary, most experts agree that the program has cost at least $20 billion while utterly failing to clean up most hazardous waste sites. Explains Carol Browner, Administrator of the Environmental Protection Agency (EPA), "A lot of time and money is taken up with companies suing each other over how much they owe to clean up a particular site."

Congress authorized $15.2 billion for CERCLA's trust fund during the 1990 budget deal. Of this amount, $10.8 billion was spent by the federal government through fiscal 1994. The amount spent by companies and individuals to comply with CERCLA has raised the overall price tag by many billions of dollars.

Much of this cost has been spent not on cleanup, but on legal fees and other "transaction costs." And while program costs have ballooned, cleanups have proceeded slowly. Only 291, or about 24 percent, of the 1,238 "worst" hazardous waste sites—sites placed on the National Priorities List (NPL)—have been cleaned up. To clean up the remaining sites already on the NPL will cost the EPA alone an estimated $40 billion. A similar amount is estimated to be spent for state sites. If the cost of cleaning up federal facilities is added to the equation, the total could exceed $750 billion, or about $7,800 per household.

SUPERFUND'S FUNDAMENTAL FLAWS

Superfund's combination of heavy cost and modest impact results from four fundamental problems with CERCLA:

- As written, the law is designed to reduce the level of chemicals in the ground, not reduce risks to people;
- The program's unreasonable standards require that hazardous waste sites be cleaned up to a pristine level no matter what the location, resulting in a poor return on the investment;
- Hazardous wastes by their very nature are local problems, but listings, cleanup decisions, and program authority are run by the federal bureaucracy; and
- The law's retroactive, strict, joint and several liability scheme results in endless legal battles as potentially liable parties fight EPA, then each other, and then their insurers to shoulder the costs of cleanup.

Each of these characteristics drives up costs while blunting the program's effectiveness. The reauthorization legislation must correct these flaws.

CLEAN UP SPILLS OR PROTECT PUBLIC HEALTH

Flaw #1: Cleaning up chemicals takes priority over protecting health.

The most important question that should be asked in the reauthorization process typically has not been asked in Superfund debates: What should be the basic intent of the law? CERCLA was enacted amid fears that some Americans were at risk from hazardous wastes. Yet the law was not designed specifically to reduce risk to the public. Rather, the goal was cleaning up spilled chemicals.

To be sure, cleaning up chemicals does help protect public health if the sites being cleaned would expose the public to real risks. But if there is no threat of risk to the public—and there would be none without a land use change—then money is squandered that could have been used elsewhere to save more lives. . . .

While it would seem prudent to protect against such potential occurrences as the likely loss of clean aquifers or threats to human health, actions to avoid hypothetical threats to life should not have the same claim on already scarce resources. Unfortunately, because Superfund is designed principally to clean up chemicals, the program costs a great deal but achieves little improvement in public health.

A Waste of Money

Flaw #2: Overly stringent cleanup standards divert resources from more effective uses.

The extremely high cost of cleanup is closely related to the question of reducing risk. For the average site, the estimated cost is about $25 million, even though, as a category, hazardous waste sites are considered by EPA to be a medium to low health risk. The cost is so unnecessarily high because of overly stringent and inflexible cleanup standards for hazardous waste sites.

In addition to requiring expensive cleanups of sites that pose little or no risk to the public, the standards require that sites posing a real risk be cleaned up to a level that goes beyond what is needed to reduce the risk to acceptable levels. For instance, polluted soil at industrial park sites must be cleaned up to a level that makes it essentially clean enough to eat. But while children may ingest contaminated dirt accidentally, or even purposefully, few are likely to be found playing in the dirt at an industrial park.

Cleanup standards should reflect the real risk to a person likely to be found at the site based on expected exposure, not on some hypothetical and very unlikely scenario. An industrial park does not need to be as pristine as a playground.

In other instances, no risk exists unless there is some change in how the land is used. Often, the owners have no plans for development or other change in current use. Yet rather than simply require a deed restriction that forbids change unless the site is cleaned up to a level appropriate for any proposed change in use, the Superfund program requires the potentially responsible party (PRP) to clean up the site even if it poses no real risk as currently used. . . .

Local, Not Federal Control

Flaw #3: The law does not permit local problems to be handled locally.

Perhaps the most puzzling feature of Superfund is that it is a federal program. Indeed, CERCLA is not delegated to the states, unlike all other major environmental laws. But hazardous waste sites are, at their core, a local problem. If there is any federal environmental law for which a legitimate case can be made for devolving authority back to the states, it is CERCLA. Contrary to the usual argument put forth by those who oppose shifting to state jurisdiction—that states cannot be trusted to protect their population—state and local officials are far more accountable for local risks under their control than are distant federal officials. It is highly unlikely that states would neglect public health and safety concerns, especially given the general awareness and level of knowledge that exist today.

To deal with any possible problems associated with devolution, certain safeguards could be built into the system. For instance, emergency removals that target sites posing immediate and substantial risks could continue at the federal level. Since the EPA has extensive experience and is relatively efficient at emergency response to sites that are truly hazardous, it could maintain the capacity to conduct emergency responses. This is not to say that the EPA emergency program is without its critics, but experience with this program generally has been better than with other parts of Superfund. Of course, safeguards would be needed to prevent abuse of what constitutes an emergency and to curtail the well-known problem of "bureaucratic creep."

There is another very important reason for returning authority over hazardous waste sites to the states: The Tenth Amendment to the Constitution reserves to the states all powers not explicitly granted to the federal government. The federal government should assume responsibility only for issues that are national in scope and interest. Federal involvement in the cleanup of local hazardous waste sites does not meet this standard. In fact, it is hard to imagine anything better suited to the exercise of state leadership and authority than local problems associated with the cleanup of hazardous waste sites.

Endless Legal Battles

Flaw #4: The determination of liability leads to endless legal battles.

One of the most damning, though inaccurate, criticisms of Superfund is that 80 percent of the money has gone to lawyers rather than for cleanup. This figure, commonly cited in the press, overstates the litigation costs severalfold, but the substance of the criticism is correct: Superfund's liability scheme encourages litigation and other transaction costs. Further, it has led to numerous "horror stories" of individual citizens and small businesses subjected to millions of dollars in potential liability for what most people would consider innocent acts, some of them committed generations ago. For instance, Russ Zimmer was named a PRP for a Superfund site at a battery cracking plant in Torrington, Wyoming. His contribution to the problem? He sold a bag of dog food in 1977 and a bag of seed in 1984, and took a third-party check as payment for the items. Since the checks had been issued by the now-bankrupt company that had owned the battery cracking plant, Zimmer was sued as a PRP by another company caught in CERCLA's liability scheme. Zimmer decided, on advice of counsel, that he should settle to avoid even more in legal costs. He agreed pay $3,500.

One of the major causes of such legal woes is the concept of joint and several liability, under which even minor contributors to a problem (for example, someone who tosses a car battery into a landfill) can be held responsible for the entire multi-million dollar cost of cleanup.

HOW THE CLEANUP MONEY IS SPENT

Breakdown of the average cost of $25 million per Superfund site for companies with annual revenues of less than $100 million.

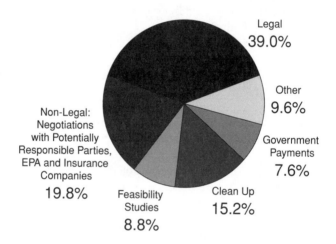

Legal
39.0%

Other
9.6%

Non-Legal: Negotiations with Potentially Responsible Parties, EPA and Insurance Companies
19.8%

Government Payments
7.6%

Feasibility Studies
8.8%

Clean Up
15.2%

RAND Corporation, 1993.

The concept is designed to find somebody—anybody—to pay for cleanup of hazardous waste, even if the "responsible" person has only a tiny part in creating the waste problem. In practice, this means that a firm with "deep pockets" ends up paying for the cleanup of wastes created by someone else. Naturally, any person or institution identified as a PRP has a strong incentive to show that others are liable as well, in order to spread the costs among more PRPs. Litigation is virtually assured.

Faced with large and uncertain liability, PRPs typically protect themselves by hiring lawyers. Thus, rather than quick payments from "polluters," the liability scheme encourages legal gridlock, cleanup delay, and enormous legal costs. A recent report estimated total government, PRP, and insurer transaction costs of about $1.3 billion at NPL sites. Although EPA Administrator Carol Browner has admitted the need for reform, she also claims that "a lot of good has come from the program. The polluter-pays concept that was first adopted in the Superfund law has changed the way businesses deal with their waste." But, Superfund does not adopt the "polluter-pays" system. Joint and several liability merely guarantees that people like Russ Zimmer pay for cleanup instead of those who are really responsible. In short, it works against the idea of making polluters pay the costs that their actions impose on others. . . .

HOW TO FIX SUPERFUND

Superfund is broken. If it is to be repaired, it must be completely overhauled. Federal agencies, state governments, municipalities, and the private sector must be relieved of the ill-conceived, unnecessary, and expensive burdens of the program. Specifically, the reformed CERCLA should at least:

1. *Turn authority for current NPL sites over to the states—with the necessary funding.* To qualify for funds, states should be required to meet two conditions.

- Prioritize sites and expenditures based on real risks.
- Eliminate joint and several liability completely . . . for sites cleaned with federal funds.

2. *Allow states to use funds for any of their most serious hazardous waste sites, not just NPL sites transferred from the federal government.* These conditions would have to apply, or states would be tempted simply to spend the money on sites for which they were responsible rather than on those that are most dangerous. . . .

If substantive reforms of this nature are not enacted, Superfund will continue to be an ineffective and expensive drain on the American economy. But if proper reforms are instituted, Congress can achieve its original goal of protecting Americans from hazardous wastes at reasonable cost.

"The culprit is pollution, not the [Superfund] program. And, in many instances, the states aided and abetted the pollution with inadequate regulation, lax permits and slow enforcement."

THE FEDERAL GOVERNMENT SHOULD MANAGE THE SUPERFUND PROGRAM

Velma M. Smith

Superfund is a federal program that began in 1980 to clean up hazardous waste sites in the United States. In the following viewpoint, Velma M. Smith argues that although the Environmental Protection Agency—the organization responsible for overseeing the Superfund program—is slow to move hazardous waste dumps off the list of contaminated sites, it does a better job of cleaning up the sites than the states would do. The federal government should continue to control Superfund, she maintains, because lax enforcement of pollution laws by the states led to the creation of many of the sites that must now be cleaned up. Smith is the executive director of Friends of the Earth, an international environmental advocacy group.

As you read, consider the following questions:

1. According to Smith, what is the Superfund program's most glaring flaw?
2. How should the Superfund program be reformed, in Smith's opinion?
3. In the author's opinion, why is the state of Colorado partially responsible for the Superfund site at the Summitville gold mine?

After 15 years of trying to clean up the nation's worst hazardous-waste sites (since 1980), it appears that Superfund has disappointed nearly everyone: businesses who sit in the shadow of enormous liabilities, citizens frustrated with explanations of health risks that raise more questions than they answer and congressional critics looking for straightforward, swift cleanups.

Some critics want to dismantle the program's liability scheme. Others would "speed up" remedies by dramatically lowering the standards by which a site is judged to be "clean." And others would simply hand the messes—and a share of federal money—back to the states.

A COMPLEX PROBLEM

In all likelihood, these options would accomplish nothing more than taking a beleaguered federal program out of the headlines for a time. And each step would repeat the program's original and most glaring flaw: underestimating and simplifying the scope and complexity of the nation's hazardous-waste problems.

In the 1970s the country was shocked by the discovery of chemicals in residential neighborhoods. Living in the chemical age was one thing; tracking dioxin into the house with the mud on your shoes was something else.

At the time it seemed that groundwater contamination would be encountered only on occasion. The slow-moving nature of groundwater and the capacity of chemicals to bind to soil were offered as solace to those who feared widespread contamination. Public officials fully expected instances of chemical-soaked soil and tainted water to be aberrations.

Expectations were wrong, but the shock has worn off. The national press no longer flocks to communities where people find they are sitting on toxic chemicals; there are simply too many.

The original list of 406 Superfund sites has grown dramatically. On this infamous list are sites that fit the image of Superfund—abandoned dumps with corroded barrels and chemicals oozing through the ground. Other sites are old but still operating facilities; nearly half are active when they are "listed." Superfund deals not only with the mistakes of the past but also with the mistakes we still are making.

THE WRONG MEASURE OF SUCCESS

It is the Environmental Protection Agency's inability to move sites off the list quickly that fuels the fervor to remake or simply rescind the federal program. Many view the size of the queue and the extended period before a site is "finished" as compelling evidence of federal ineptitude.

Numbers and time offer clear yardsticks, but they are the wrong measures for judging this important federal program. Hindsight shows that Republican Sen. William Cohen of Maine was right in 1980 when he cautioned that it would "take generations for our resources to recover" from the "problem of toxic wastes poisoning our environment."

Clearly, our Superfund is more than one of sheer numbers. By definition, sites on the National Priorities List are the worst of the worst. The scoring system used to rank sites considers proximity to people, impact on drinking water and other factors, and the EPA has determined that, at more than 80 percent of the sites, "actual human exposure and/or actual contamination of a sensitive environment has taken place."

Superfund critics argue that these exposures are inconsequential, but a number of scientific studies have documented troubling links between proximity to hazardous waste sites and a host of health problems from cancer to birth defects. Evidence of harm is not ironclad, but for residents of Columbia, Miss., with skin lesions, for five Houston mothers of children with a rare facial deformity and for families who have lost children to leukemia in communities from Woburn, Mass., to Riverside, Calif., the compelling need is not for more data but for more protection.

Ironically, while regulators are roundly criticized for being too "conservative" with risk assessment, predictions of how fast, how deep or how far pollutants move have not been "conservative" enough. Over the years, scientists have found that a variety of chemicals move through the environment in unexpected ways, that soil and groundwater contamination are less the exception than the norm.

Cleanup Is Difficult

Contamination cases also are far more difficult to remedy than once theorized. In the San Gabriel Valley of California, for example, "plumes" of contaminated groundwater cover 40 square miles and reach depths of 700 feet. The implications are serious; there are about 275 public-water supply wells in the area. For the Valley, an ad hoc strategy of coping with the contamination has not worked. As decisions have been made to close wells or pump from new "clean" wells, the movement of the plume has changed, and contaminated water has been drawn into previously uncontaminated areas. Simply shutting off the taps and letting Mother Nature dilute the problem won't work;

the water basin serves as a drinking-water source to more than a million people.

Californians are frustrated that the EPA has made so little progress, but the valley's troubles simply cannot be "solved" for many, many years. Like many Superfund problems, this one will demand long-term management, ongoing cleanup efforts and monitoring and oversight for generations to come.

Superfund can and should do better: Site investigations and interim controls should be initiated earlier; studies should be more thoughtful and thorough; communities should be kept in the know, not in the dark; state regulators should be consulted up front; and better data on the performance and the cost of cleanup technologies should be disseminated.

The EPA can go a long way in improving its performance, but the public must face the fact that it may not be able to do much better in terms of "finishing" cleanup and checking sites off the list. And no one should imagine that the states, which vary widely in their ability to manage cleanup and their culpability in creating the problems in the first place, would do a significantly better job. Simply returning the list to each state with a check from the federal treasury and best wishes for success would likely magnify rather than alleviate the problems.

THE SUMMITVILLE GOLD MINE

Consider the cleanup of a 1,000-acre gold mine in Summitville, Colo., first proposed for the Superfund list in 1993. Even then the restoration project had been costing the federal taxpayers up to $25 per minute, not to buy a pristine cleanup, but simply to keep thousands of gallons of cyanide waste from running into the headwaters of the Rio Grande. Summitville comes to the Superfund program thanks to a series of errors by the state of Colorado. "Everything that could go wrong at Summitville did go wrong," according to Mike Long of the Colorado Division of Minerals.

In 1984 the now-bankrupt Summitville Consolidated Mining Co. submitted a permit application for a gold-mine operation of enormous proportions. With what has been described as "unseemly haste," regulators approved construction of the "leach pad" that would hold more than 9 million tons of ore to be pumped through with a cyanide solution to remove the gold.

The state, which halved the regulatory program's funding in 1986, responded slowly to problems. Like their counterparts elsewhere, the Colorado regulators tried for a long time to "work it out" with the company; a large investment and a number of jobs were at stake, after all. In July 1991 and again in July 1992, regulators and the company set down agreements on improvements to control the operations. In December 1992, the company declared bankruptcy. At that point, the state called the EPA, and using the emergency authorities of Superfund the agency began working to keep cyanide and acid wastes from washing down the valley. The cleanup costs are projected to run to $170 million. That sum will stabilize and control; it won't restore Summitville to its pre-mining state.

STRENGTHEN SUPERFUND

Superfund needs to be strengthened, not weakened. Congress should support legislation that would:
- Continue to make polluters, not taxpayers, pay for cleanup of toxic waste sites.
- Clean up hazardous waste sites completely, safely, and permanently.
- Prevent future pollution by expanding public reporting of toxic chemical emissions and initiating reporting on toxic chemical use.
- Give citizens a greater role in the cleanup of their neighborhoods.
- Repeal the current exemption from liability enjoyed by oil companies.

U.S. Public Interest Research Groups, "Make Polluters Pay," October 30, 1998.

Myths to the contrary, "pristine" is seldom part of the Superfund lexicon. Large numbers of sites already have been "written off" for sensitive future uses such as schools and homes, even where those sites are neighbors to other homes, playgrounds and churches. Superfund stands as testament to the fact that Americans throw away far more than plastic bottles and newspapers; in many communities, land and the land-use options of future generations have been used up by pollution.

Not all Superfund stories are as dramatic as Summitville. Still, too many of the stories repeat a pattern of slow-moving, ineffective and shortsighted environmental management and community complaints about fumes, discharges or dumping unheeded by state regulators. Despite disaffection with the EPA, many citizens would despair entirely if cleanup responsibilities reverted back to the states.

A TOTAL DISASTER

Says the Environmental Health Network's Linda King: "Giving the Superfund program over to the states would be a total disaster for residents of poor rural states. Some of those who live near Superfund sites have borne the brunt of pollution, and their lives and livelihoods have become political footballs tossed about by uncaring state politicians." According to King, intervention from Washington is critical.

Superfund can be improved. But laying all blame for today's Superfund mess on the EPA is a bit like blaming the Resolution Trust Corp. for savings-and-loan failures. And handing the job back to the states—without a careful consideration of each state's capacity and inclination to clean up—would rival an appointment of Charles Keating to head a commission on banking reform. [Keating was convicted in 1993 of federal racketeering, conspiracy, and securities fraud in a scandal that led to the collapse of Lincoln Savings and Loan.]

The course that was charted in 1980 is not wrong; it is, however, longer and much more difficult than predicted. Our sobering Superfund experience stands as an indictment—not of an inept Washington bureaucracy but of ineffective controls on polluting. The culprit is pollution, not the program. And, in many instances, the states aided and abetted the pollution with inadequate regulation, lax permits and slow enforcement. They should be first in line for criticism, not cleanup dollars. They certainly shouldn't be given yet another opportunity to sit by and merely watch while the pollution problems worsen.

> "Governors and legislators have added a variety of independently supported state structures that outlast any political regime and act as a counterweight to political moves to cut environmental protection."

THE STATES SHOULD ENFORCE POLLUTION LAWS

Mary Graham

Mary Graham argues in the following viewpoint that most pollution problems are now local problems and state governments are the best resource for controlling and monitoring pollution. Therefore, she contends, the states should be given the authority to regulate and police pollution within their borders. She asserts that as attitudes toward the environment have changed, businesses are no longer given the green light to pollute. Graham is a fellow at Harvard University's Kennedy School of Government.

As you read, consider the following questions:

1. In Graham's opinion, why is the idea of "race to the bottom" an outdated idea?
2. Why are environmental costs rarely a determining factor in siting a new business location, according to the author?
3. What is the best way to mediate between national priorities and state differences, in Graham's view?

Excerpted from Mary Graham, "Why States Can Do More: The Next Phase in Environmental Protection." Reprinted with permission from *The American Prospect*, January/February 1998. Copyright 1998 The American Prospect, PO Box 383080, Cambridge, MA 02138. All rights reserved.

W hen Congress laid the foundation for today's environmental regulation in the 1970s, it was an article of faith
that states inevitably cut corners in conservation and pollution
control in order to attract business. Only the federal government, the argument went, had the political clout and national
reach to prevent a state "race to the bottom."

Seemingly, there is new support for this view. Not long ago,
the press carried lurid reports of hog wastes washing down Virginia's Pagan River toward the Chesapeake Bay. The federal Environmental Protection Agency (EPA) sued Smithfield Foods, Inc.,
the East Coast's largest producer of pork products, accusing Virginia Governor George Allen of lax enforcement of national water
pollution laws. The state's failure "could create 'pollution havens'"
and "lead to a shift of manufacturing and jobs that would penal-
ize the conscientious states," the *New York Times* editorialized.

AN OUTDATED IDEA

But "race to the bottom" is far too simplistic a notion to describe
state environmental politics in the 1990s. The idea is outdated
for three reasons. First, state policies have been transformed by
three decades of national regulation and increased public con-
sciousness, which give pollution and conservation issues a larger
voice at the state level, quite independent of current federal action. Second, evidence is by now overwhelming that businesses
rarely decide where to locate or expand based on the strength or
weakness of state environmental laws. Third and most important,
public attitudes have changed. After 30 years of government action and scientific progress, state officials, business executives,
and voters find that some environmental measures have economic value. Assuring safe drinking water, limiting growth, or
preventing sewage from polluting public beaches, for example,
may produce benefits that residents and businesses can appreci-
ate and are sometimes willing to pay for. More often, though,
support for environmental protection follows, rather than pre-
cedes, state economic success. Today, states compete to gain prosperity in a fast-changing economy. In that race, some states lead
in economic performance *and* environmental protection, while
others lag behind in both.

Of course, there are still tough trade-offs to be made. Hiring
inspectors to enforce pollution laws or buying land to protect a
watershed is expensive, and must vie for limited state funds
with improving schools, building roads, and paying for Medi-
caid and welfare. Nor is controversy avoidable. Environmental issues often do pit jobs against cleaner air or better conservation,

and sometimes both choices offer economic benefits. When stakes are high, business, labor, homeowners, and other groups will fight to protect their interests.

A Turning Point in Environmental Policy

The question of whether race-to-the-bottom pressures exist is important right now, because we have reached a turning point in national environmental policy where some readjustment of federal and state roles is inevitable. Thanks in part to the considerable success of national rules aimed at controlling major sources of pollution and encouraging conservation of resources on large tracts of federal land, public attention is now turning to problems that are harder to solve from Washington. The next generation of environmental policies will tackle widely scattered sources of pollution and conservation opportunities that affect farms and housing developments as well as forests and meadows.

Lessening contaminated runoff from fields and streets, the largest remaining source of water pollution, means addressing politically volatile questions of how homeowners manage their property and how farmers use their land. The considerable difficulties of monitoring pollutants from industrial and municipal pipes pale before the intricacies of controlling septic tank maintenance, grazing practices, and fertilizer use. And the complexities of conserving resources on public land are small compared to the challenge of protecting ecosystems that are a patchwork of farms, housing developments, and forests. "Most of the problems remaining now . . . are site-specific, varying from area to area and requiring tailored controls at the regional, state, or local level for effective mitigation," the EPA Science Advisory Board concluded in 1990. The National Academy of Public Administration has suggested that state and local governments might be better at making decisions about chemical contaminants in drinking water, and about when, where, and how to clean up hazardous wastes.

Both Republicans and Democrats are calling for new approaches to traditional environmental problems, to combat the cost and inflexibility of "command-and-control" regulation. EPA administrator Carol Browner has suggested that federal rules be supplemented with negotiated solutions, market incentives, and industry standards, and she predicted recently that in ten years environmental standards will be set "place by place," rather than "pollutant by pollutant."

Because most of our daily attention is drawn to hard-fought battles at the perimeter of government authority, it is easy to forget that we have witnessed an exceptional event in the last three decades: the successful introduction of a new theme in national policy. Mainstream Democrats and Republicans agree that air pollution, water pollution, and other environmental problems that cross state lines should continue to be controlled by federal rules.

During the fight in 1997 over tightening standards for soot and smog, no voices were raised in favor of giving up federal rules altogether. In fact, the battle itself was a dramatic illustration of the need for national standards, since it pitted upwind midwestern states protecting their coal-burning power plants against downwind northeastern states concerned about their partially imported dirty air. Likewise, it took an amendment to the Clean Air Act in 1990 to balance the interests of midwestern utilities that burn high-sulfur coal against the needs of northeastern states and Canada where their emissions contribute to acid rain. . . .

STATES HAVE CHANGED

While congressional gridlock has slowed efforts to update national policy, the state role in environmental protection has fundamentally changed. Many aspects of pollution control and conservation have been assimilated into state and local politics, as they have been into national politics. In a particularly lucid essay in *Environmental Policy in the 1990s: Reform or Reaction?*, political scientist Barry Rabe of the University of Michigan notes that about 70 percent of important environmental legislation enacted by the states now has little or nothing to do with national policy, and that only about 20 percent of the $10 billion that states now spend annually on environment and natural resources comes from Washington.

State and local governments are responsible for nearly all the enforcement of national environmental laws, and continue to dominate decisions in areas like land use and waste disposal where Congress has not acted. Most states have replaced a traditional public health department orientation with single agencies that manage environmental protection, and three years ago those agency heads formed an Environmental Council of the States to encourage state initiative and federal flexibility. Echoing the National Environmental Policy Act, most states require assessment of environmental impact of important public actions, and a combination of state and federal rules require public hearings before permits are granted to industry to discharge pollutants in the air, in the water, or on land. . . .

BUSINESSES HAVE CHANGED

Industry has now been living with strong federal and state environmental regulation for nearly three decades. Today, environmental costs rarely are a determining factor in business location, in part because national laws have encouraged all businesses to improve their practices. In the 1970s, sudden new government requirements with short deadlines frequently called for large, unplanned investments that were costly to industry. Now, however, pollution control and conservation costs generally have become a small and predictable element of business operating expenses. Even for chemical and petroleum industries, annual pollution control expenses run less than 2 percent of sales. Capital expenditures for pollution abatement vary widely from one industry to another, ranging from 2 percent of total capital costs for machinery and 3 percent for electronics to 13 percent for chemical industries and 25 percent for petroleum and coal. Even when substantial, though, those costs are usually dwarfed by labor, real estate, transportation, energy, and tax considerations in relocation decisions, according to surveys of corporate executives. . . .

THE STATES CAN PROTECT THEIR ENVIRONMENT

When increasing air pollution created public concern in the 1960s, many states and localities enacted meaningful laws. But Congress responded to the pleas of auto manufacturers and electric utilities to stave off state control by interposing an ineffectual federal bureaucracy. The resulting 1965 and 1967 Clean Air Acts both flopped. . . . The 1970 Clean Air Act gave the EPA power over the states. The EPA claims great success in reducing pollution from stationary sources. But according to a Brookings Institution study, the states' 1960s laws reduced these pollutants three times more than those enacted in the 1970s, when the EPA was in charge.

David Schoenbrod, *Wall Street Journal*, May 8, 1997.

These days, when a company relocates, it may get an unexpected dose of state scrutiny of its environmental practices as part of the bargain, even from a state that is competing hard to gain business. That's what happened when the Intel Corporation, producer of the Pentium computer chip, announced a decision in 1993 to build the world's most advanced semiconductor plant in Rio Rancho, New Mexico. The company's move provoked local protests that directed national attention to pollution and conservation issues in an industry generally seen as environmentally friendly, and induced Intel to cut proposed water use and to spend unanticipated millions on pollution control.

PUBLIC ATTITUDES HAVE CHANGED

Further, the idea that states routinely minimize environmental protection to attract business is outdated because many state politicians, business executives, and voters now understand that some cleanup and conservation measures have economic value, particularly if they contribute to critical infrastructure, attract skilled workers, or satisfy the needs of particular businesses. Voters have shown that they are sometimes willing to pay for environmental protection if benefits fall within state borders and will be experienced soon. . . .

New Jersey, a small, densely populated state traditionally heavily dependent on chemical manufacturing and petroleum-related industry, has been a leader in environmental protection. Over the years, the state's voters supported the nation's first mandatory recycling law, restrictions on wetland development more severe than those included in federal rules, and a right-to-know law that was the model for the federal Toxic Release Inventory, telling people about dangerous chemicals discharged by industry.

Sometimes, state initiatives have been inspired by sudden threats to public health. After a 50-mile slick of floating garbage washed up on Massachusetts beaches in 1987, the state initiated a $200 million program to clean up coastal areas. In response to evidence that a quarter of drinking-water wells were contaminated with pesticides, Iowa legislators increased pesticide registration fees and fertilizer taxes to fund research on how to reduce chemical use in agriculture. . . .

WHAT TO DO?

Giving up on the simplistic theme of a "race to the bottom" among states to minimize environmental protection opens the way for considering harder questions. How much flexibility should states have to make choices about environmental measures? How can national priorities not in the interest of any one state best be advanced? How should chronic inequities among states be dealt with? A number of initiatives already underway suggest constructive approaches.

State Flexibility. Setting clear national goals and giving states as much flexibility as possible in how to carry them out is the best way to mediate between national priorities and state differences. The National Academy of Public Administration has suggested a strategy of "differential oversight" to concentrate federal monitoring and enforcement wherever states, taking their various paths in pursuit of prosperity, are weakest. Supplementing standards with wider use of market incentives, negotiated solutions, and business self-certification can also broaden local choices while respecting national priorities. And as knowledge improves, state progress should be judged by changes in pollution and conservation, rather than by numbers of inspections and permits. All of this is, of course, much easier said than done. After 30 years of efforts and billions of dollars spent, the United States still does not have a reliable system of measuring trends in environmental conditions that could be a basis for setting national goals and marking progress toward them.

Information as Regulation. Requiring that the public receive detailed information, interpreted objectively, can create incentives for business and governments to limit pollution. The 1996 amendments to the Safe Drinking Water Act, passed by the 94th Congress after two years of acrimony, require local water systems to notify customers once a year about bacteria and chemicals in tap water as one way of encouraging careful monitoring. And a number of states require electric utilities to report on environmental factors. Using "Surf Your Watershed," the newest EPA internet site, anyone who enters a zip code can now get specific information about pollution sources, water quality, and drinking water sources. These information requirements follow the example of the Toxic Release Inventory, a provision added to federal law in 1986 that requires companies to report on their discharges of toxic substances. . . .

It would be a mistake to let the fears of the 1970s dominate action into the next century. The "race to the bottom" is a powerful idea that resonated with sudden changes in environmental requirements during the 1970s. It has little bearing on the challenges of the 1990s when environmental costs are a relatively small portion of business expenses, state governments have constructed means of weighing their constituents' environmental needs in public decisions, and public understanding has improved. After nearly 30 years, environmental protection has been assimilated into the political system, where it will continue to evolve in thousands of separate national, state, local, and private actions.

"Governors are engaged in a frantic race to lure dirty industries within their borders by promoting their states as refuges from environmental law enforcement."

THE FEDERAL GOVERNMENT SHOULD ENFORCE POLLUTION LAWS

Robert F. Kennedy Jr.

In the following viewpoint, Robert F. Kennedy Jr. argues that the states should not be given control over enforcing and regulating pollution laws. This power should remain with the federal government. The states cannot be trusted to maintain the same high standards of pollution control as the federal government, Kennedy asserts, as is evident in the fact that nearly half the states have given polluting corporations immunity from prosecution. Kennedy is an environmental lawyer and the coauthor of a book on environmental policy, *The Riverkeepers*.

As you read, consider the following questions:

1. Who, according to Kennedy, are the most hostile to prosecuting polluters?
2. How many polluters did the state of Virginia prosecute during 1996, as cited by the author?
3. In what way do the residents of the Hudson River Valley live with toxic waste, according to Kennedy?

Reprinted from Robert F. Kennedy Jr., "Why Our Waters Need Washington," *George*, December 1997, by permission of the author.

The fall of 1997 marks the twenty-fifth anniversary of the Clean Water Act. Passed in 1972 over Nixon's veto, the law promised that all American waterways would be swimmable and fishable by 1983. In 1995, Capitol Hill Republicans began a well-publicized effort to bring back the good old days when the Cuyahoga River burned, Lake Erie was proclaimed dead, and inoculations were recommended for anyone who swam, fished, or boated on the Mississippi. By rewriting the Clean Water Act, conservative Republicans hoped to remove federal wetlands protection and many of the restrictions on dumping sewage and toxic chemicals into our waterways. Fortunately, public outrage derailed the "dirty water bill."

But now the Clean Water Act and other environmental laws are under attack again, this time in the state capitols. GOP governors are engaged in a frantic race to lure dirty industries within their borders by promoting their states as refuges from environmental law enforcement. Predictably, the states most hostile to prosecuting polluters—South Carolina, Louisiana, Alabama, and Virginia—tend to be the most polluted.

By gutting already weak and underfunded enforcement programs, so-called law-and-order Republicans are now making the feeble records of their Democratic predecessors seem positively green. In 1996, the state of Virginia, under Governor George Allen, prosecuted only one polluter the entire year (down from an already meager 30 cases in 1989) and collected a grand total of $4,000 in fines. Under Louisiana's Republican governor, Mike Foster, the number of prosecutions has dropped 54 percent since the previous administration. Water-pollution enforcement has nearly collapsed under GOP governors in Rhode Island, Montana, and South Carolina and has been seriously hobbled in Alabama, Mississippi, and Pennsylvania. A poll of Virginia environmental employees found that 57 percent feared reprisal by their superiors if they made a decision that might upset a polluter.

Even in New York, where Governor George Pataki has established some bona fide green credentials in other areas, enforcement cases have dropped 45 percent since 1992. In 1996, an upstate New York developer's attorney told me, "From our clients' point of view, it's a wonderful thing. It's my job to tell them that for the moment, there is no environmental enforcement in New York state. It's had an obvious impact on their conduct. Everyone knows now that you can get away with anything."

To make matters worse, since 1994, 24 states have passed laws offering corporations immunity from prosecution as long as they disclose their illegal activities. Now there's a smart idea. Why not go one step further and let criminals go free in exchange for confessions?

POLLUTION DOES NOT RECOGNIZE STATE BORDERS

Environmental policy cannot simply be delegated to the states, as the libertarians pretend. Air and water recognize no state boundaries. Supporters of the House bill repeatedly said that Maryland could protect the Chesapeake Bay without Federal regulation if the state was genuinely concerned about its recreation and fishing industries. But Maryland would have no way to stop polluted water originating in Virginia or Pennsylvania.

Sherwood Boehlert, *New York Times*, July 5, 1995.

Republicans claim they want to dismantle the federal laws in the name of states' rights or community control. In New York's Hudson Valley, we hear those phrases as euphemisms for corporate control. Before federal environmental laws went into effect in the '70s, corporations in New York, including General Electric, routinely threatened to move plants to New Jersey if New York tried to uphold its environmental standards. Hudson Valley residents are still living with the toxic legacy of GE's blackmail: a \$280 million bill to clean up General Electric's polychlorinated biphenyls (PCBs), contamination that has devastated the river's 350-year-old commercial fishery and cost the area hundreds of jobs. More seriously, elevated levels of PCBs have been found in the breast milk of lactating women between Oswego and Albany. And everyone in the Hudson Valley has PCBs in their flesh and organs.

Environmental laws like the Clean Water Act are intended to protect public health and safety, and the environment. They emerged from a democratic movement to wrest control from corporations, which could easily dominate state political landscapes. With the establishment of these federal laws, small communities could, for the first time, stand up to the big corporations and demand that they comply with federal standards. The Republican attack on environmental enforcement threatens to lead us back to the epoch when companies like GE could bully communities to ransom their children's future for the promise of a few years of prosperity. Republicans and their corporate sponsors need to be told that dirty water doesn't taste any better when served up by state officials.

"Citizens will recognize that the common law, bolstered by local regulation, can protect the environment more effectively and fairly than can congressional statutes and bureaucratic regulations."

THE LEGAL SYSTEM SHOULD REGULATE POLLUTION

Roger E. Meiners and Bruce Yandle

In the past, disputes regarding pollution were handled by the court system. Those found guilty of polluting were subject to strict penalties and thus many potential polluters took measures to avoid being sued. In the following viewpoint, Roger E. Meiners and Bruce Yandle argue that the fear of lawsuit was an effective deterrent to pollution. The authors believe that when pollution ceased to be a concern of the courts and was instead turned over to government agencies, the effectiveness of pollution control was diminished. In the authors' opinion, a return to court-regulated pollution control would protect the environment and its residents better than any laws and regulations. Meiners is a professor of law and economics at the University of Texas in Arlington. Yandle is alumni professor of economics and legal studies at Clemson University.

As you read, consider the following questions:

1. What was the court's decision in *Carmichael v. City of Texarkana*, as cited by the authors?
2. What are some problems with the common law system as a means of protecting the environment, according to the authors?

Reprinted from Roger E. Meiners and Bruce Yandle, "Curbing Pollution Case-by-Case," *PERC Reports*, June 1998, by permission of the publisher, the Political Economy Research Center, Bozeman, Montana. Endnotes in the original have been omitted in this reprint.

Unless you are well into middle age or were a precocious student, you probably have little memory of the United States without the Environmental Protection Agency (EPA) and the host of federal statutes it implements. You may not be aware of the long history of environmental controls through common-law protections. (Common law is the term we use for the legal rules and traditions that have been developed over time through court decisions.)

In the past, people who allowed something noxious to escape their control and invade the property of others could be held accountable through legal actions for trespass and nuisance. This protection was extended to water quality through riparian rights, which allow water users the right to the use and enjoyment of water. The following cases illustrate common-law protection of surface water, groundwater, and air.

CARMICHAEL VERSUS CITY OF TEXARKANA

In the late nineteenth century, the Carmichael family owned a 45-acre farm in Texas, with a stream running through it, that bordered on the state of Arkansas. The city of Texarkana, Arkansas, built a sewage system and connected numerous residences and businesses to it. The sewage collected by the city system was deposited in front of the Carmichaels' homestead, about eight feet from the state line, on the Arkansas side. The Carmichaels sued the city in federal court in Arkansas.

The court found that the cesspool was a "great nuisance" that polluted the stream on the Carmichaels' property, "depositing the foul and offensive matter . . . in the bed of said creek on plaintiffs' land and homestead continuously. . . ." The court said that this cesspool deprived the family of the "use and benefit of said creek running through their land and premises in a pure and natural state as it was before the creation of said cesspool. . . ."

The claims for damages were awarded, as was the plaintiffs' suit for an injunction against the cesspool. The court cited a leading text on the law of torts:

> If a riparian proprietor has a right to enjoy a river so far unpolluted that fish can live in it and cattle drink of it and the town council of a neighboring borough, professing to act under statutory powers, pour their house drainage and the filth from water-closets into the river in such quantities that the water becomes corrupt and stinks, and fish will no longer live in it, nor cattle drink it, the court will grant an injunction to prevent the continued defilement of the stream, and to relieve the riparian proprietor from the necessity of bringing a series of actions for the

daily annoyance. In deciding the right of a single proprietor to an injunction, the court cannot take into consideration the circumstance that a vast population will suffer by reason of its interference.

Judge Rogers noted: "I have failed to find a single well-considered case where the American courts have not granted relief under circumstances such as are alleged in this bill against the city. . . ."

LIABLE FOR POLLUTION DOWNSTREAM

So, long before the Environmental Protection Agency came into existence, municipalities and firms knew that if they substantially polluted their neighbors' water, they could expect to be found liable. To minimize liability, water polluters installed pollution control devices. Paper mills in Wisconsin routinely owned miles of downstream river property, knowing that otherwise they would be liable for violation of riparian rights.

Such common-law protection of water applies to all who have the right to use the water, for purposes including recreation. Those who enjoy sport fishing in England have long protected water quality through private litigation brought by angling associations.

GROUNDWATER CONTAMINATION

The primary problem caused by pollution of land is groundwater pollution due to seepage from improperly disposed of wastes. Wastes that are properly contained rarely cause harm. After Congress passed the Superfund law in 1980 (the Comprehensive Environmental Response Compensation and Liability Act) to regulate the cleanup of toxic waste sites, relatively few common-law cases have occurred. However, they show how the common law might have dealt with the problem of groundwater pollution over time.

For example, in 1981, the Illinois EPA, backed by the federal EPA, supported the right of a chemical waste landfill to remain in operation. The landfill had been built with state and federal approval, but residents of a nearby village alleged that the landfill was damaging their water supply.

The Illinois supreme court agreed. It held that the landfill was a public and a private nuisance. The village residents were there first; their right not to have their property damaged could not be stripped in favor of a "general societal" desire for a landfill. In other words, toxic landfills are legitimate, but they must be constructed so as not to impose costs on surrounding landowners who have not agreed to the intrusion. The court issued a permanent injunction against the landfill and ordered that the toxic wastes be dug up, moved, and the land restored.

AN EVOLVING VIEW

The courts' view of the standards for groundwater contamination has evolved over the decades, as the 1982 case of *Wood v. Picillo* illustrates. Neighboring property owners sued a farmer who maintained a hazardous waste dump on his property. They claimed that the dump emitted noxious fumes and polluted groundwater. The Rhode Island supreme court agreed. In doing so it overturned a 1934 decision that would have supported the defendant's position. The 1934 decision was based on the state of science at that time, when knowledge about the course of groundwater was, as the court stated in 1982, "indefinite and obscure." Since 1934, the court said:

> the science of groundwater hydrology as well as societal concern for environmental protection has developed dramatically. As a matter of scientific fact the courses of subterranean waters are no longer obscure and mysterious. . . . We now hold that negligence is not a necessary element of a nuisance case involving contamination of public or private waters by pollutants percolating through the soil and traveling underground routes.

This means that the common law now imposes strict liability (that is, liability even when there is no negligence) on polluters who cause damage to waters. This standard of care is consistent with old common-law tort rules imposing strict liability in case of hazardous materials, rules recorded in a famous 1868 British case, *Rylands v. Fletcher*. This case is often cited for restating the ancient proposition, "So use your property as not to injure your neighbor's property."

COMMON LAW AND AIR POLLUTION

The courts have long recognized common-law liability for air pollution when liability can be assigned to a polluter causing harm. An early case was *Georgia v. Tennessee Copper Co.*

The state of Georgia, on behalf of its citizens, sued two companies that operated copper smelters in Tennessee near the Georgia border. Justice Holmes noted that a public nuisance had been created because the "sulphurous fumes cause and threaten damage on so considerable a scale to the forests and vegetable life, if not to health, within [several counties in Georgia]. . . ." Defendants argued that they had recently constructed new facilities that reduced the scope of the problem, but the Supreme

Court held for Georgia. The Court gave the companies a reasonable time to build more emission-control equipment, but held that if such equipment did not reduce emissions enough to protect plant life in Georgia, the state could ask the court for an injunction to shut down the smelters.

Common Law Protects the Environment

There are no specific common laws that prohibit or limit environmental pollution. However, where specific harm can be shown, common law can be used to force a polluter to cease, repair the harm or pay damages. Specific aspects of common law that can be invoked in environmental cases include:

- *The Common Law of Trespass.* An individual or corporation can trespass on someone's property not only by physically entering the property, but by causing polluted air or water or waste to enter the property. Provided that a property owner can show that he or she was harmed, the land owner can sue a polluter.

- *The Common Law of Nuisance.* A public nuisance is created when someone acts in such a way as to disturb the public, or to limit the public's use or enjoyment of public property. Public nuisances can include smoke, noise and odor. A current example is the odor problem created by hog farms. The common law of nuisance can also be used by a private individual, if the nuisance occurs on the individual's property.

- *The Common Law of Negligence.* A person who acts negligently and thus causes harm may be held liable, if specific damages can be shown. Disposal of hazardous materials on someone else's property, especially if deliberate, may be found to involve negligence. Pesticide drift during spray operations may be negligent (as well as trespass or a nuisance).

Natural Resource Conservation and Management Program at the University of Kentucky, "The Roots of Environmental Law and Policy," November 9, 1998.

In 1915 the parties returned to the Supreme Court. The companies showed that their new, expensive equipment cut emissions by more than half. Georgia argued that this was not enough and demanded that the smelters be closed. The chief justice appointed a scientist from Vanderbilt University to spend six months, at company expense, studying the emissions and the likely effect of new controls. In the meantime, the Court ordered the companies to cut back production to reduce emissions further. Based on the evidence presented by the scientist, the companies would either be allowed to continue operation with more emission-control equipment in place, or, if that could not reduce emissions sufficiently, would have to shut down. Finally, after following the guidance of the Vanderbilt professor, the firm satisfied the plaintiffs, and the Court ended its oversight of the case.

A Few Problems

The common-law approach was not perfect. The cases reported here represent the majority view, but there was a minority view, also. Some courts would rule for polluters, holding that the economic benefit of a factory that employed many people outweighed the damage to a few property holders. Some courts held that pollution was just a fact of modern life and necessary for progress to occur. The courts were not always consistent in their decisions.

There were other problems. For one, legal action is always costly. Second, multiple polluters that each inflict low levels of damage are unlikely to be held liable—especially when the damage is shared by many. For that reason, problems with air pollution caused by automobiles cannot be handled effectively through common-law courts.

Injuries and harms that come after long gestation periods present another challenge. While parties who can show evidence of injury or imminent harm may have a common-law cause of action, efforts to obtain injunctions for speculative harms such as future cancer are not generally successful.

However, we cannot know how the law might have evolved had it not been pushed to one side by regulation. Environmental courts—ordinary courts assisted by special masters trained in environmental science—and other arrangements might well have evolved to satisfy the needs of people concerned about their environmental rights.

Time for a Return to Common Law

In our view, it is time to consider a return to the regime that served us well in the past and that has shown signs of evolving as knowledge and environmental concerns changed. The common law provides harsh penalties against firms that disregard the rights of citizens by exposing them to harms. Indeed, when real harm is inflicted, citizens get far better relief through common-law suits than they do from appeals to the Environmental Protection Agency. Eventually, we believe that citizens will recognize that the common law, bolstered by local regulation, can protect the environment more effectively and fairly than can congressional statutes and bureaucratic regulations.

"If taxing something means we get less of it, then why not tax pollution—thereby cleaning up the environment and raising new revenue at the same time?"

Businesses Should Be Taxed on Pollution

Ted Halstead

Ted Halstead argues in the following viewpoint that pollution taxes would protect the environment by encouraging businesses and individuals to generate less pollution. Furthermore, the revenue generated from pollution taxes would reduce or eliminate payroll taxes, boost wages, and create jobs, he maintains. Halstead, a senior fellow at the World Policy Institute and a Montgomery Fellow at Harvard University's Kennedy School of Government, is the founder and president emeritus of Redefining Progress, a public policy think tank in San Francisco.

As you read, consider the following questions:

1. How much new revenue would be generated by carbon-emissions permits, as cited by Halstead?
2. How would the process of implementing pollution taxes and phasing out payroll taxes make the entire tax system more progressive, in the author's opinion?
3. Which program should receive the revenues from pollution taxes, according to Halstead?

Excerpted from Ted Halstead, "Why Tax Work?" The Nation, April 20, 1998. Reprinted with permission from The Nation.

The best thing that could happen to our tax code is a substantial reduction in the payroll tax, which is now the largest and most damaging tax on working Americans. With all the partisan rhetoric about the value of hard work, it's amazing that the one tax that falls exclusively on work and wages has gotten so little attention. Yet payroll taxes impose a huge and unfair burden on low-income workers and small businesses, hindering job creation and cutting into take-home pay.

POLLUTION TAXES

Instead of taxing payrolls, America should tax pollution. Adopting market-based policies to clean up the environment could yield hundreds of billions of dollars in new public revenue. Using this money to cut existing payroll taxes would strengthen our economy, boost wages and job creation, fix our troubled tax system and protect the environment, all without raising the deficit. What more could Americans want from a tax plan? . . .

Here, then, is a proposal that could not only generate hundreds of billions of dollars in new revenue from a virtually untapped source but also encourage socially beneficial behavior among corporations and individuals. Economists tell us that we will get less of what we tax and more of what we don't. While this highlights the absurdity of taxing work, it also points to the wisdom of using the price system to solve environmental ills. If taxing something means we get less of it, then why not tax pollution—thereby cleaning up the environment and raising new revenue at the same time?

Americans are big fans of sin taxes, which explains why President Bill Clinton is proposing a new cigarette tax. It's a short leap from taxing alcohol and cigarettes to taxing pollution. Recently a group of 2,500 U.S. economists (including eight Nobel laureates) issued a statement supporting the use of new environmental levies to cut existing taxes. Meanwhile, the British Treasury has announced plans to eliminate payroll taxes at the bottom of the wage scale and to introduce new pollution taxes, which should raise enough revenue to offset the payroll tax cuts. The logic is impeccable: Replacing harmful and unproductive taxes with pollution taxes would lead to a stronger economy and a cleaner environment.

TRADABLE PERMITS

One of the most pressing environmental problems of our day is global warming, which is caused by heat-trapping greenhouse gases (like carbon dioxide) emitted in the burning of fossil fuels. President Clinton, who has joined the scientific mainstream in declaring that global warming must be addressed, recently proposed a system of tradable pollution permits to reduce U.S. greenhouse-gas emissions. This kind of market-based approach has much in common with a carbon tax—which is a levy on the carbon content of fossil fuels—only it doesn't sound like one. In essence, a tradable-permit system would cap the nation's total carbon emissions and allow individual companies to buy and sell the limited number of pollution rights.

DRI/McGraw-Hill estimates that using a permit approach to reduce our emissions to 1990 levels—which is slightly less ambitious than Clinton's proposal—would generate more than $140 billion per year in new revenue. If this were applied to payroll tax relief, it would allow the first $10,000 of wages to be exempted, on both the individual and business sides. And other environmental levies—including taxes on water pollution and the use of virgin materials—could gradually be introduced to reduce payroll taxes even further.

This type of shift in the source of public revenue would make the whole tax system more progressive in several ways. Exempting the first $10,000 of wages from payroll taxes would benefit the neediest the most. And unlike Social Security payroll taxes, which apply disproportionately to wage earners and labor-intensive companies, new pollution levies would also apply to trust-funders and capital-intensive firms. Even if companies tried to pass the higher fuel costs on to consumers, this would be more than offset by the fact that oil prices recently hit a fourteen-year low.

THE BEAUTY OF SHIFTING TAXES

Neither issue—reducing payroll taxes or imposing pollution charges—has a winning political constituency on its own. Combined, they just might. Whereas payroll tax cuts have never had a chance without an acceptable source of replacement revenue, the benefits of earmarking new pollution levies for the Social Security trust fund may be compelling enough to win the support of retirees. And while past proposals to raise pollution costs have pitted environmentalists against labor and unified industry in opposition, linking such proposals to payroll tax cuts could lead to just the opposite: a newfound alliance between enviros and labor and a splitting of the corporate community.

The beauty of shifting from payroll taxes to pollution levies to fund Social Security is that there's something in it for almost everyone—everyone, that is, except the coal and oil industries (whose sales would go down as the costs of pollution went up and investments in energy conservation suddenly became more attractive). Working Americans would see their wages, benefits and job prospects increase as the costs of labor declined. Social Security recipients would have the comfort of knowing their retirement is being secured in a way that is good for the country, its economy and their grandchildren. And small businesses, which have long pointed to payroll tax cuts as their number-one legislative priority, should also be strong supporters.

TAX SHIFTS FROM WORK AND INVESTMENT TO ENVIRONMENTAL DAMAGE

Country, Year Initiated	Taxes Cut On	Taxes Raised On	Revenue Shifted[1] (percent)
Sweden, 1991	Personal income	Carbon and sulfur emissions	1.9
Denmark, 1994	Personal income	Motor fuel, coal, electricity, and water sales; waste incineration and landfilling; motor vehicles	2.5
United Kingdom, 1996–97	Wages	Landfilling	0.2

[1]Expressed relative to tax revenue raised by all levels of government.

David Malin Roodman, Worldwatch Paper 134, "Getting the Signals Straight: Tax Reform to Protect the Environment and the Economy," May 1997.

The more surprising allies may well come from big industry. After all, service- and knowledge-based industries, which tend to be labor intensive but consume relatively little energy, now constitute 80 percent of the economy. It stands to reason that these would reap significant benefits from reduced labor costs in exchange for increase pollution fees. Even the auto industry is beginning to see there is money to be made in the efficiency revolution: General Motors recently announced its support for higher energy taxes. . . .

Substituting pollution levies for payroll taxes could symbolize a much broader message: that Americans should be able to keep more of the fruits of their own toil, but that they ought to pay for the costs they impose on current and future generations. Isn't it time we bring our economic incentives back into line with our shared values?

> "Making the polluter pay should not entail trying to eliminate the generation of wastes. . . . It means . . . ensuring that polluters pay for the costs of the harms they inflict upon others."

"POLLUTER PAYS" POLICIES ARE UNFAIR AND INEFFECTIVE

Jonathan H. Adler

Although the "polluter pays" principle—which requires polluters to pay for damages caused by their harmful waste products—is valid, it rarely achieves the desired effect of reducing pollution, contends Jonathan H. Adler in the following viewpoint. Many of the fines imposed on polluters are due to violations of complex rules and not because of illegal emissions, he maintains. Furthermore, the notion of imposing taxes on pollution would serve only to generate revenue for the government and do little to compensate those who are harmed by the pollution. The "polluter pays" principle can be effective if it is reformed so that victims are compensated by polluters for the damages caused by pollution, Adler argues. Adler is the director of environmental studies at the Competitive Enterprise Institute, a public policy think tank that advocates limited government and free-market principles.

As you read, consider the following questions:

1. In Adler's opinion, what is pollution?
2. What was the Exxon *Valdez*'s crime, according to Adler?
3. According to the author, why are English rivers relatively free of pollution?

Reprinted from Jonathan H. Adler, "Making the Polluter Pay," *The Freeman*, March 1995, by permission of *The Freeman*. Endnotes in the original have been omitted in this reprint.

The experience of the past few decades indicates that "pollution control" is often a pretext by which the federal government regulates the minutiae of each and every industrial process and economic transaction. Much of this so-called pollution control is done in the name of the "polluter pays" principle. This principle, which is intuitively sensible, was trumpeted by early environmentalists as a means to discourage environmental harms.

The "polluter pays" rhetoric is still often used, and most Americans probably think that current environmental policies make polluters pay. In truth, however, this approach is seldom embodied in American environmental laws.

Rarely are particular polluters forced to pay for actual damage caused. For example, when Congress enacted Superfund, the federal program to clean up hazardous waste, "polluter pays" was used to justify generic taxes on producers of materials (chemicals and oil) that ended up in waste dumps. Even if companies had acted responsibly—even if none of their materials or products ended up at waste sites—and they had caused no damage, they had to pay the tax if they happened to produce certain materials. Superfund is a policy under which polluters and nonpolluters alike are forced to pay exorbitant sums.

POLLUTER PAYS PRINCIPLE NEEDS REFORM

The polluter pays principle is valid, but it needs to be better understood and, ultimately, to be reinstated under institutional arrangements that make it effective and fair. To begin with, one must recognize that emissions per se are not pollution. Pollution is the imposition of a harmful waste product or emission onto the person or property of another without that person's consent; it is a "trespass" under the principles of common law. If the trespass is so minor that it creates no impact or inconvenience for the property owner, it will normally be tolerated. Otherwise, it will likely result in legal action of some kind.

The generation of a waste, in and of itself, does not necessarily harm other people or their property. Not every emission, waste, discharge, or industrial by-product is pollution. Thus there is no reason for government policy to discourage waste per se. Yet environmental regulators are eager to adopt "pollution prevention," "waste reduction," and "toxics-use reduction" schemes. Such programs completely miss the point. They tend to move away from any true concern for limiting pollution, and from holding polluters accountable for the damages that they cause.

Current environmental policy rarely focuses on harm. Indeed, sometimes it doesn't even focus on pollution at all! Much of the time the emphasis is on compliance with byzantine rules and requirements. Fines are levied not when the property of another is contaminated, but when a permit is improperly filed, or a waste-transport manifest is not completed in line with the demands of regulatory officials. The Environmental Protection Agency itself has observed that under current law "a regulated hazardous waste handler must do hundreds of things correctly to fully comply with the regulations, yet doing only one thing wrong makes the handler a violator." Environmental rules are now so complex that only 30 percent of corporate counsels believe that full compliance with environmental laws is actually possible, according to a survey conducted by the *National Law Journal*.

THE EXXON VALDEZ CASE

Even when harm occurs as a result of pollution, the "polluter pays" principle is routinely violated. Consider the case of the Exxon *Valdez*. In 1989, an oil tanker ran aground because its captain was drunk, and over 300,000 barrels of crude poured into the water of Prince William Sound, causing significant, though not permanent, environmental disruption. Few people are aware that the crime for which Exxon was punished was killing migratory birds without a permit. Extensive shorelines were covered in oil, and the government prosecuted Exxon for not having permission to go hunting!

Exxon was subject to civil suits from those, such as local fishermen, who claimed damage from the spill. However, much of the money that Exxon was forced to pay did not go to alleged victims of the spill. Exxon paid $125 million in fines to the federal government and the state of Alaska. In addition, Exxon was forced to pay $900 million into a fund to be doled out by government officials for environmental projects, habitat protection, and scientific research, among other things. In May 1994, $38.7 million of this money was used to create a new state park.

Exxon was under tremendous political pressure to restore the "public" shoreline so it engaged in a costly, and extensive, cleanup operation. Much of the cleanup was unnecessary—nature has its own methods of cleaning up spills of natural substances like oil—and in some cases the extensive beach cleaning actually caused harm. So, not only was Exxon prosecuted on generic offenses against "public" goods rather than for specific harms to specific parties, but the politicization of the spill resulted in a thoughtless policy response. Had a similar spill occurred in a more private setting—if, for example, a tanker truck had overturned, spilling onto private properties—the owners of the affected properties would have had clear, direct recourse. Additionally, they would have had a tangible incentive to ensure that any cleanup or remediation was a proper way to address the problem at hand.

There was no means for affected citizens to hold Exxon directly responsible for much of the actual damage caused to the Alaskan shoreline. The Alaskan coast had no private owners, stewards, or protectors who could seek redress or ensure that cleanup dollars were well spent, as they could if that oil had spilled into someone's backyard. The only direct payments made by Exxon to those actually harmed were to fishermen and Alaska natives who claimed damages from a temporary decline in the salmon and seal harvest.

If we truly want polluters to pay, there need to be private property owners that can defend threatened or harmed resources. Ownership of ecological resources can serve as a deterrent against causing harm against others, in the same manner that private property provides such incentives in other areas. Private ownership also provides tangible incentives for better stewardship.

Polluters such as Exxon should be held responsible, not for violating a bureaucratic proscription about the hunting of birds or for having harmed some "public" resource, but because they harmed someone else's person or property, and they have no right to do that. Moreover, any restitution should be paid to those harmed, not simply to a government agency that proclaims it will spend the money in the public interest.

MAKING POLLUTERS PAY

A fishing club in England, the Pride of Derby Angling Club, demonstrates how property rights can prevent stream pollution. In England, clubs own the right to fish along some rivers and they protect their "beats" from pollution. In 1948, several fishing club members joined to form the Anglers' Co-operative Association (ACA). The association won a major case soon thereafter, known as the Pride of Derby case. Upstream polluters were required to stop polluting, and pay damages and legal costs, since their pollution threatened the fishery. The ACA has helped fishing clubs pursue injunctions against upstream pollution ever since. To date, the ACA has been involved in over 1,500 cases, including several against municipal water authorities.

This ability of private parties to restrain upstream polluters is rarely available in the United States. Historically, some communities and individuals did obtain traditional common law remedies for water pollution. However, many such actions have since been preempted by the federal Clean Water Act. Under the Clean Water Act, politically preferred polluters are treated more favorably than others. Municipal polluters face cleanup goals that are often less stringent than those of industrial polluters, and their cleanup schedules are far more lenient. Yet, to the rivers and fish, pollution is pollution.

CITIZEN LAWSUITS

This problem of unequal treatment is compounded by the prevalence of citizen suit provisions in the Clean Water Act and other environmental laws. Although it may sound good to allow any citizen or citizen group to force the government to enforce pollution laws (and to allow the citizen or group to recoup legal costs), what it means is that special interest groups can effectively determine the enforcement priorities of government agencies. Many of the environmental organizations that engage in citizen suits have an anti-business bias. As a result, private industry is subject to more legal actions than either agricultural activities or governmental facilities, even though both of the latter are greater sources of water pollution. Indeed, between 1984 and 1988, environmentalist citizen suits against private industry were more than six times as common than suits against governmental facilities. "There are no environmental reasons why environmental groups would display such a pronounced preference for proceeding against corporate polluters," notes Michael Greve of the Center for Individual Rights.

Many environmental groups have found that citizen suits can be a lucrative source of revenue. There is something profoundly unjust about limiting the legal recourse of persons harmed by polluting activities, as the politicization of pollution control has done, while at the same time encouraging the use of citizen suits by organizations with no stake in the resources they claim to be protecting.

AIR POLLUTION

Another example of failure to make polluters pay is the case of air pollution. It is well established that a small fraction of automobiles are responsible for the vast preponderance of auto-related emissions. Indeed, over half of all auto emissions are generated by only ten percent of the cars on the road. This means that for every ten cars, the dirtiest one pollutes as much as the other nine. But federal officials insist upon imposing significant costs on the owners of all cars through "clean fuel" requirements, periodic emissions inspections, and the like, in order to meet federal air quality standards. If emission reductions are necessary in some regions to protect human health (an arguable proposition), targeting the dirtiest portion of the automobile fleet would reduce pollution more efficiently and more equitably. Indeed, if air-sheds were managed privately, one would expect this sort of approach to emissions reductions.

THE TROUBLE WITH POLLUTER-PAYS

The polluter-pays methodology assumes that scientists and policymakers can accurately determine who pollutes what and how much. This is a task of such breathtaking complexity that it is impractical, if not impossible, to rely on it as a substitute for political decisions.

In the end, the polluter-pays methodology is a scorched-earth tactic that will just make the resolution of controversial environmental issues more difficult, leading to more protracted and messy stalemates. That's bad news for the environment and for the economy.

Martha Edenfield, *Florida Business Insight*, March/April 1998.

The broad approaches (which I call "drift-net" approaches) achieve pollution reductions more through their scope than their efficiency and tend to produce environmental improvements at the expense of innocent individuals who have not contributed to environmental harm. Environmental protection and simple justice are better served when pollution reduction efforts focus on the true sources of pollution, and ensure that it is the polluters that pay for the damages caused.

DO POLLUTION TAXES WORK?

There is one other approach that appears to embody the "polluter pays" principle: the imposition of emission taxes. This idea is generally associated with the economist A.C. Pigou, who argued that pollution taxes would force offending industries to "internalize" the costs they were imposing on others.

But there are several problems with this approach. First, such taxes would be used to enrich government coffers, not to compensate those who were harmed by the pollution. It is one thing for the state to decide disputes and ensure that polluters make restitution to those whom they have harmed. It is another thing for the state to identify polluting activities and use pollution taxes as a source of general revenue. The former is in accord with common law principles of justice; the latter encourages the continued growth of the regulatory state.

The second problem is that the state is in no position to assess the actual costs imposed by pollution. Pollution taxes enacted through the political process are likely to reflect political priorities rather than environmental ones. The federal gasoline tax, for example, is often defended as a "polluter pays" approach because oil exploration, refining, and use all have environmental impacts. However, a tax on gasoline is a poor proxy for taxing environmental impacts—the same gallon of gasoline will produce different levels of emissions in different vehicles. And special-interest pleading ensures that certain types of fuels and fuel additives receive special exemptions from the tax.

In fact, pollution tax schemes almost inevitably rely upon some proxy for pollution that can be taxed. It is far easier to levy a tax on an easily measurable factor, such as use of a resource or aggregate emissions, than it is to try and measure the impact on people—yet it is the impact on people and the environments that they are concerned about that should matter. Using tax mechanisms in place of common law principles, no matter how well intentioned the policy, is a "polluter pays" approach that is destined to fail.

In sum, making the polluter pay should not entail trying to eliminate the generation of wastes and other by-products of a modern industrial society. Nor does it mean regulating every emission, every industrial process, indeed every aspect of economic life. It means focusing environmental protection efforts on the greatest sources of harm and ensuring that polluters pay for the costs of the harms they inflict upon others. This goal can be best accomplished through a decentralization of environmental policy and a greater reliance upon common law remedies. Central government dictates are not up to the task.

PERIODICAL BIBLIOGRAPHY

The following articles have been selected to supplement the diverse views presented in this chapter. Addresses are provided for periodicals not indexed in the *Readers' Guide to Periodical Literature*, the *Alternative Press Index*, the *Social Sciences Index*, or the *Index to Legal Periodicals and Books*.

John H. Cushman Jr.	"Courts Expanding Effort to Battle Water Pollution," *New York Times*, March 1, 1998.
Alfred M. Duda	"Downwind, Downstream," *Forum for Applied Research and Public Policy*, Fall 1998.
Ann Glumac	"Minnesota Makes Strides in Controlling Pollution," *Forum for Applied Research and Public Policy*, Winter 1996.
Stephanie B. Goldberg	"Let's Make a Deal," *ABA Journal*, March 1997.
James M. Inhofe	"Federal Regulators Are Set to Choke Local Economies for the Sake of Marginal Health Benefits," *Insight*, April 28, 1997. Available from 3600 New York Ave. NE, Washington, DC 20002.
Roger E. Meiners and Bruce Yandle	"Get the Government Out of Environmental Control," *USA Today*, May 1996.
John G. Mitchell	"Our Polluted Runoff," *National Geographic*, February 1996.
Manesh Podar and Richard M. Kashmanian	"Charting a New Course," *Forum for Applied Research and Public Policy*, Fall 1998.
Christopher R. Powicki	"Treating Industrial Wastewaters Naturally," *World & I*, January 1998. Available from 3600 New York Ave. NE, Washington, DC 20002.
David Malin Roodman	"Subsidizing Pollution," *Humanist*, May/June 1997.
David Schoenbrod	"State Regulators Have Had Enough of the EPA," *Wall Street Journal*, May 8, 1997.
John Schofield	"Pollution for Sale," *Maclean's*, March 23, 1998.
Dwight G. Smith	"Unseen Causes of Unclean Waters," *World & I*, July 1996.
Jerry Taylor	"State Cleanups Are Cheaper and Faster," *Insight*, May 1, 1995.

FOR FURTHER DISCUSSION

CHAPTER 1

1. Lynn Landes argues that hazardous waste should be stored until it can be safely recycled. Based on your readings in this book, do you think storing hazardous waste until it can be recycled is preferable to burning or burying it? Why or why not? Is storing it a practical and effective solution to the disposal of hazardous waste? Support your answer with examples from the viewpoints.

2. Dick Russell argues that the seas and coastlines and their accompanying marine life are dying from pollution runoff. Gary Turbak maintains, however, that the oceans, rivers, and harbors are the cleanest they've been in decades. What evidence does each author offer to support his argument? Which argument seems stronger? Why?

CHAPTER 2

1. Both Stephen Lester and the American Council on Science and Health cite studies to support their view on whether or not dioxin is a health hazard to humans who have been exposed to the chemical. Based on your reading of the viewpoints, do you think dioxin is a health hazard? Support your answer with examples from the viewpoints.

2. Richard Wiles, Kert Davies, and Christopher Campbell claim that the additive effect of pesticides must be considered when determining a level for a "safe" dose of toxic chemicals. How does Michael Fumento respond to this assertion? Which argument is strongest? Explain your answer.

CHAPTER 3

1. Doug Bandow asserts that there is no reason to recycle because recycling wastes resources by creating more pollution and uses more energy than manufacturing items from raw materials. How does Allen Hershkowitz respond to Bandow's contention? Which argument do you think is strongest? Support your answer with examples from the viewpoints.

2. According to John Tierney, recycling glass, plastic, metal, and paper is not worth the time, effort, or money required to separate and collect the recyclable materials. Based on your readings of the viewpoints, do you agree with Tierney's conclusions? Why or why not?

3. The National Research Council received funding for its re-

search on sewage sludge from agencies such as the Environmental Protection Agency, the U.S. Department of Agriculture, the U.S. Food and Drug Administration, the Association of Metropolitan Sewerage Agencies, several municipal water districts, and companies that develop technology for treating sewage sludge. Does knowing that the report received grants from these different organizations influence your assessment of its argument? Explain your answer.

CHAPTER 4

1. Kenneth Green argues that the new air quality regulations proposed by the Environmental Protection Agency are too costly for businesses and industries. How does Carol M. Browner respond to his concerns? In your opinion, should the expected costs of implementing the new air quality standards affect the decision to revise the regulations? Why or why not?

2. Brian Tokar contends that allowing companies to buy and sell pollution credits is essentially giving them the "right" to pollute. Do you agree with his argument? Support your answer with examples from the viewpoints.

3. Drew Kodjak and Richard de Neufville and his colleagues debate the advantages and disadvantages of owning and driving electric vehicles. On which points do they agree? On which do they differ? Based on your reading of the viewpoints, which argument is stronger? Explain your answer.

CHAPTER 5

1. The Superfund law requires that all sites contaminated with hazardous wastes be cleaned to a "pristine" condition. John Shanahan does not believe it is necessary to clean an industrial park site to the same standard as a residential housing site if a restriction is noted on the deed forbidding a change in land use. Do you agree or disagree with Shanahan's view? Explain your answer.

2. John Shanahan, Velma M. Smith, Robert F. Kennedy Jr., and Mary Graham all argue over whether the federal government or the states should have more control over environmental laws and regulations. Based on your reading of the viewpoints in this chapter, who should have responsibility for ensuring hazardous waste sites are cleaned up? Who should determine and enforce water and air quality standards? Support your answers with examples from the viewpoints.

ORGANIZATIONS TO CONTACT

The editors have compiled the following list of organizations concerned with the issues debated in this book. The descriptions are derived from materials provided by the organizations. All have publications or information available for interested readers. The list was compiled on the date of publication of the present volume; the information provided here may change. Be aware that many organizations take several weeks or longer to respond to inquiries, so allow as much time as possible.

American Council on Science and Health (ACSH)
1995 Broadway, 2nd Fl., New York, NY 10023-5860
(212) 362-7044 • fax: (212) 362-4919
e-mail: acsh@acsh.org • website: http://www.acsh.org

ACSH is a consumer education consortium concerned with, among other topics, issues related to the environment and health. The council publishes *Priorities* magazine and position papers such as "Global Climate Change and Human Health" and "Public Health Concerns About Environmental Polychlorinated Biphenyls."

Canadian Centre for Pollution Prevention (C2P2)
100 Charlotte St., Sarnia, ON, N7T 4R2, CANADA
(800) 667-9790 • fax: (519) 337-3486
e-mail: c2p2@sarnia.com • website: http://c2p2.sarnia.com

The Canadian Centre for Pollution Prevention is Canada's foremost pollution prevention resource. It offers easy access to national and international information on pollution and prevention through a search service, hard copy distribution, an extensive website, on-line forums, publications, and customized training. Among their publications are the *Practical Pollution Training Guide* and *At the Source*, C2P2's quarterly newsletter.

Cato Institute
1000 Massachusetts Ave. NW, Washington, DC 20001-5403
(202) 842-0200 • fax: (202) 842-3490
e-mail: cato@cato.org • website: http://www.cato.org

The institute is a libertarian public policy research foundation dedicated to limiting the role of government and protecting individual liberties. It researches claims of discrimination and opposes affirmative action. The institute publishes the quarterly magazine *Regulation* and the bimonthly *Cato Policy Report*. It disapproves of EPA regulations, considering them to be too stringent. The Cato Institute publishes over one hundred papers dealing with the environment with titles ranging from "Why States, Not EPA Should Set Pollution Standards" and "The EPA's Clean Air-ogance."

Earth Systems
PO Box 1413, Charlottesville, VA 22902
(804) 293-6398 • fax: (804) 296-9825
e-mail: www@earthsystems.org • website: http://earthsystems.org

Earth Systems is a nonprofit organization that develops, compiles, categorizes, and delivers environmental education and information resources to the world at large. It offers a virtual library with an index of over nine hundred on-line environmental resources and EcoTalk, a mailing list devoted to helping nonprofit environmental organizations find solutions to problems.

Environmental Industry Associations (EIA)
4301 Connecticut Ave. NW, Suite 300, Washington, DC 20008
(202) 244-4700 • fax: (202) 966-4818
e-mail: eia@envasns.org • website: http://www.envasns.org
Affiliated with the National Solid Wastes Management Association and the Waste Equipment Technology Association, EIA represents about two thousand businesses that manage solid, hazardous, and medical wastes; manufacture, distribute, and service waste equipment; and provide environmental management and consulting services. It publishes the newsletter *Infectious Wastes News* and offers several pamphlets and profiles on various waste management issues.

Environmental Protection Agency (EPA)
401 M St. SW, Washington, DC 20460-0001
(202) 382-2090
website: http://www.epa.gov
The EPA is the federal agency in charge of protecting the environment and controlling pollution. The agency works toward these goals by assisting businesses and local environmental agencies, enacting and enforcing regulations, identifying and fining polluters, and cleaning up polluted sites. It publishes the monthly *EPA Activities Update* and numerous periodic reports.

Friends of the Earth
1025 Vermont Ave. NW, Suite 300, Washington, DC 20005
(202) 783-7400 • fax: (202) 783-0444
e-mail: foe@foe.org • website: http://www.foe.org
Friends of the Earth is a national advocacy organization dedicated to protecting the planet from environmental degradation; preserving biological, cultural, and ethnic diversity; and empowering citizens to have an influential voice in decisions affecting the quality of their environment. It publishes the quarterly *Friends of the Earth Newsmagazine*.

Heritage Foundation
214 Massachusetts Ave. NE, Washington, DC 20002-4999
(800) 544-4843 • (202) 546-4400 • fax: (202) 544-6979
e-mail: pubs@heritage.org • website: http://www.heritage.org
The Heritage Foundation is a conservative think tank that supports free enterprise and limited government. Its researchers criticize EPA overregulation and believe that recycling is an ineffective method of dealing with waste. Its publications, such as the quarterly *Policy Review*, include studies on the uncertainty of global warming and the greenhouse effect.

INFORM

120 Wall St., 16th Fl., New York, NY 10005-4001
(212) 361-2400 • fax: (212) 361-2412
website: http://www.informinc.org

INFORM is an independent research organization that examines the effects of business practices on the environment and on human health. The collective goal of its members is to identify ways of doing business that ensure environmentally sustainable economic growth. It publishes the quarterly newsletter INFORM *Reports* and fact sheets and reports on how to avoid unsafe uses of toxic chemicals, protect land and water resources, conserve energy, and safeguard public health.

Natural Resources Defense Council (NRDC)

40 W. 20th St., New York, NY 10011
(212) 727-2700
e-mail: proinfo@nrdc.org • website: http://www.nrdc.org

The Natural Resources Defense Council is a nonprofit organization that uses law, science, and more than four hundred thousand members nationwide to protect the planet's wildlife and wild places and to ensure a safe and healthy environment for all living things. NRDC offers hundreds of publications, including *Cool It: Eight Great Ways to Stop Global Warming* and *After Silent Spring: The Unsolved Problems of Pesticide Use in the United States.*

Political Economy Research Center (PERC)

502 S. 19th Ave., Suite 211, Bozeman, MT 59718
(406) 587-9591 • fax: (406) 586-7555
e-mail: perc@perc.org • website: http://www.perc.org

PERC is a nonprofit research and educational organization that seeks market-oriented solutions to environmental problems. Areas of research covered in the PERC Policy Series papers include endangered species, forestry, fisheries, mines, parks, public lands, property rights, hazardous waste, pollution, water, and wildlife. PERC conducts a variety of conferences, offers internships and fellowships, and provides environmental education materials.

Worldwatch Institute

1776 Massachusetts Ave. NW, Washington, DC 20036-1904
(202) 452-1999 • fax: (202) 296-7368
e-mail: worldwatch@worldwatch.org
website: http://www.worldwatch.org

Worldwatch is a nonprofit public policy research organization dedicated to informing policymakers and the public about emerging global problems and trends and the complex links between the world economy and its environmental support systems. It publishes the bimonthly *World Watch* magazine, the Environmental Alert series, and several policy papers.

BIBLIOGRAPHY OF BOOKS

Frank Ackerman	*Why Do We Recycle? Markets, Values, and Public Policy.* Washington, DC: Island Press, 1996.
Nicholas A. Ashford and Claudia S. Miller	*Chemical Exposures: Low Levels and High Stakes.* New York: Van Nostrand Reinhold, 1997.
Dennis T. Avery	*Saving the Planet with Pesticides and Plastic: The Environmental Triumph of High-Yield Farming.* Indianapolis: Hudson Institute, 1995.
Harvard Ayers, Charles E. Little, and Jenny Hager	*An Appalachian Tragedy: Air Pollution and Tree Death in the Eastern Forests of North America.* San Francisco: Sierra Club Books, 1998.
Joanna Burger	*Oil Spills.* New Brunswick, NJ: Rutgers University Press, 1997.
Steve Coffel	*Encyclopedia of Garbage.* New York: Facts On File, 1996.
Theo Colborn, Dianne Dumanoski, and John Peterson Myers	*Our Stolen Future: Are We Threatening Our Fertility, Intelligence, and Survival? A Scientific Detective Story.* New York: Plume, 1997.
Carlos D. DaRosa, James S. Lyon, and Philip M. Hocker	*Golden Dreams, Poisoned Streams: How Reckless Mining Pollutes America's Waters, and How We Can Stop It.* Bozeman, MT: Mineral Policy Center, 1997.
Edward F. Dolan	*Our Poisoned Waters.* New York: Cobblehill Books, 1997.
Paul Ehrlich and Anne Ehrlich	*Betrayal of Science and Reason: How Anti-Environmental Rhetoric Threatens Our Future.* Washington, DC: Island Press, 1996.
Dan Fagin and Marianne Lavelle	*Toxic Deception: How the Chemical Industry Manipulates Science, Bends the Law, and Endangers Your Health.* Secaucus, NJ: Birch Lane Press, 1996.
Michael Fumento	*Polluted Science: The EPA's Campaign to Expand Clean Air Regulations.* Washington, DC: AEI Press, 1997.
Lois Marie Gibbs	*Dying from Dioxin: A Citizen's Guide to Reclaiming Our Health and Rebuilding Democracy.* Boston: South End Press, 1995.
Joshua Karliner	*The Corporate Planet, Ecology, and Politics in the Age of Globalization.* San Francisco: Sierra Club Books, 1997.
J.S. Kidd and Renee A. Kidd	*Into Thin Air: The Problem of Air Pollution.* New York: Facts On File, 1998.
Allan Mazur	*A Hazardous Inquiry: The Rashomon Effect at Love Canal.* Cambridge, MA: Harvard University Press, 1998.

Christina G. Miller

Air Alert: Rescuing the Earth's Atmosphere. Boston: Atheneum, 1996.

Michael Sanera and Jane S. Shaw

Facts Not Fear: A Parent's Guide to Teaching Children About the Environment. Washington, DC: Regnery, 1996.

Michael Shnayerson

The Car That Could: The Inside Story of GM's Revolutionary Electric Vehicle. New York: Random House, 1996.

Marvin S. Soroos

The Endangered Atmosphere: Preserving a Global Commons. Columbia: University of South Carolina Press, 1997.

John Stauber and Sheldon Rampton

Toxic Sludge Is Good for You! Lies, Damn Lies, and the Public Relations Industry. Monroe, ME: Common Courage Press, 1995.

Sandra Steingraber

Living Downstream: An Ecologist Looks at Cancer and the Environment. New York: Addison-Wesley, 1997.

John Wargo

Our Children's Toxic Legacy: How Science and Law Fail to Protect Us from Pesticides. New Haven, CT: Yale University Press, 1998.

INDEX